Illuminate
Publishing

15 99

WJEC
Biology
for A2 Level

Study and
Revision Guide

Neil Roberts

Published in 2017 by Illuminate Publishing Ltd, P.O Box 1160,
Cheltenham, Gloucestershire GL50 9RW

Orders: Please visit www.illuminatepublishing.com
or email sales@illuminatepublishing.com

British Library Cataloguing in Publication Data

A catalogue record for this book is available from the British Library
ISBN 978-1-908682-53-6

Printed in England by 4Edge Ltd., Hockley, Essex.

2nd impression: 5.17

The publisher's policy is to use papers that are natural, renewable and recyclable
products made from wood grown in sustainable forests. The logging and manufacturing
processes are expected to conform to the environmental regulations of the country of
origin.

Every effort has been made to contact copyright holders of material reproduced in this
book. If notified, the publishers will be pleased to rectify any errors or omissions at the
earliest opportunity.

This material has been endorsed by WJEC and offers high quality support for the
delivery of WJEC qualifications. While this material has been through a WJEC quality
assurance process, all responsibility for the content remains with the publisher.

WJEC examination questions are reproduced by permission from WJEC.

Editor: Geoff Tuttle
Cover and text design: Nigel Harriss
Text and layout: Kamae Design, Oxford

Acknowledgments

For Isla and Lucie.
The author would like to thank the editorial team at Illuminate Publishing for their support
and guidance.

About the author

Neil Roberts is a former Head of Biology and has over 20 years teaching experience in
universities, schools and colleges in England and Wales, and was an experienced principal
examiner for a major awarding body. In 2009, he was awarded with a fellowship of the Royal
Society of Biology.

Contents

How to use this book

Knowledge and Understanding

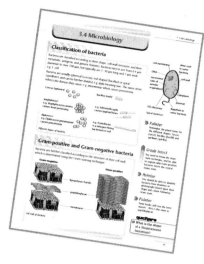

The first section of the book covers key knowledge required for the examination. There are notes on:

- Unit 3, Energy, Homeostasis and the Environment
- Unit 4, Variation, Inheritance and Options

There is a practical assessment in A2, but you also need to be able to answer practical based questions in the exam.

Examples and exam tips have been included in the guide to help you prepare.

You will also find:

- **Key terms**: many of the terms in the WJEC specification can be used as the basis of a question, so we have highlighted those terms and offered definitions.

- **Quickfire/Extra questions**: are designed to test your knowledge and understanding of the material as you go along.

- **Pointer and Grade boost**: offer extra examination advice to improve your exam technique and raise your exam performance.

- **Additional practice questions**: feature at the end of the book providing you with further practice at answering questions with a range of difficulty.

Approximately 10% of your marks will come from the assessment of your mathematical skills. Help is provided throughout the guide, including worked examples.

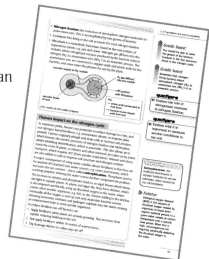

Exam Practice and Technique

The second section of the book covers the key skills for examination success and offers you examples based on suggested model answers to possible examination questions. First, you will be guided into an understanding of how the examination system works, an explanation of Assessment Objectives and how to interpret the wording of examination questions and what they mean in terms of exam answers.

This is followed by a selection of examination and specimen questions with actual student responses. These offer a guide as to the standard that is required, and the commentary will explain why the responses gained the marks that they did.

It is a good idea to split the course down into manageable chunks, complete revision notes as you go along, and have a go at as many questions as you can. The real key to success is practice, practice, practice past paper questions, so I advise that you look at www.WJEC.co.uk for sample papers and past papers. A level is such a big jump from GCSE; you really need to start working for the exams from day 1!

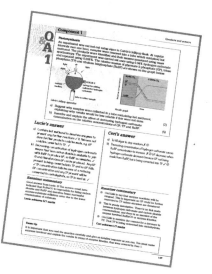

Unit 3   Knowledge and Understanding

2. Photosynthesis uses light energy to synthesise organic molecules.
pages 10–21

3. Respiration releases chemical energy in biological processes.
pages 22–28

1. The importance of ATP.
pages 8–9

4. Microbiology.
pages 29–33

Energy, Homeostasis and the Environment

5. Population size and ecosystems.
pages 34–47

8. The nervous system
pages 65–77

6. Human impact on the environment.
pages 48–55

7. Homeostasis and the kidney
pages 56–64

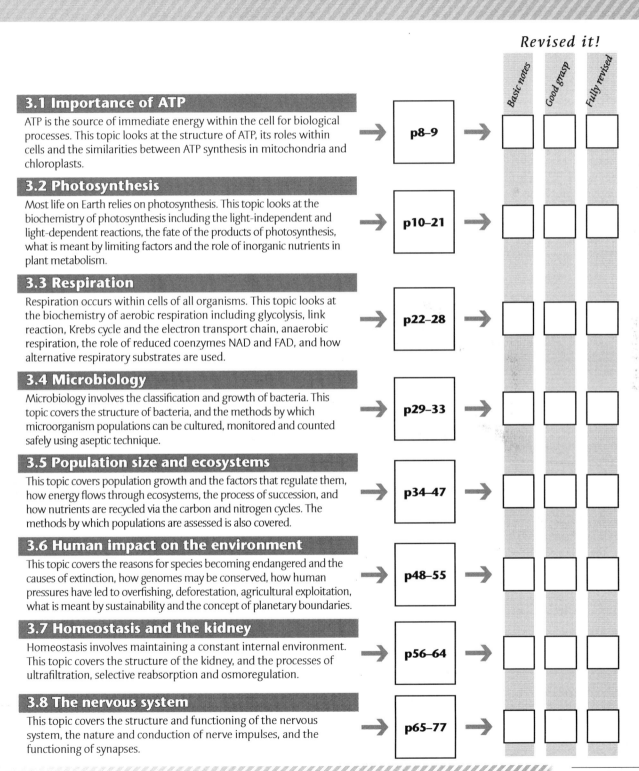

3.1 Importance of ATP

Structure and roles of ATP

Adenosine triphosphate (ATP) belongs to a group of molecules called nucleotides. It is made from ribose and adenine (which are collectively referred to as ribulose) and three phosphates. It is the universal energy carrier (used in all reactions in all organisms), and releases energy in small quantities (30.6 kJ per mol) via a one-step reaction when the high energy bond between the second and third phosphate group is broken. This hydrolysis reaction is hydrolysed by the enzyme ATPase.

Link See pages 51–52 in AS Study and Revision Guide.

Grade boost

Revise your work on ATP.

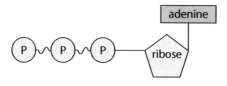

The structure of ATP

When ATP is hydrolysed, it provides energy for a wide range of processes including: protein synthesis, muscle contraction, DNA synthesis, active transport and mitosis.

The mitochondria and chloroplast membranes

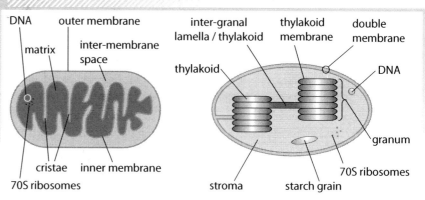

Basic structure of a mitochondrion   *Basic structure of a chloroplast*

During photosynthesis and respiration, ATP is made when protons are pumped across the membranes using energy from electrons to create an electrochemical or proton gradient. When the protons flow back through the stalked particles down the concentration gradient, by a process known as **chemiosmosis**, ATP synthetase phosphorylates ADP into ATP. In chloroplasts this occurs on the

Key Term

Chemiosmosis: the flow of protons down an electrochemical gradient, through ATP synthetase, coupled with the synthesis of ATP from ADP and a phosphate ion.

thylakoid membranes, whereas in mitochondria it occurs on the inner membrane or cristae. The electrons pass from the proton pumps to a terminal electron acceptor: in mitochondria this is oxygen, in chloroplasts it is the coenzyme NADP or chlorophyll. The mechanisms by which ATP is generated in photosynthesis and respiration are explained in more detail later (see pages 10 and 22).

(see pages 10 and 22)

quickfire

① How is chemiosmosis different from osmosis?

Comparison of ATP synthesis in mitochondria and chloroplasts

Feature	Mitochondria	Chloroplasts
Mechanism	Uses energy carried by electrons to pump protons across the membrane, they then flow back through stalked particles	Uses electron energy to pump protons across the membrane, which then flow back through stalked particles
Enzyme involved	ATP synthetase	ATP synthetase
Proton gradient	From inter-membrane space to matrix	From thylakoid space to stroma
Site of electron transport chain		Thylakoid membrane
Co-enzyme involved	FAD, NAD	
Terminal electron acceptor		NADP and H^+ (non-cyclic photophosphorylation) and chlorophyll$^+$ (cyclic photophosphorylation)

quickfire

② Complete the table to compare synthesis of ATP in mitochondria and chloroplasts.

Types of phosphorylation

Phosphorylation is the addition of a phosphate group or ion to a molecule. In respiration and photosynthesis ADP is the molecule most often phosphorylated, but other molecules can be phosphorylated, e.g. glucose in glycolysis, forming glucose diphosphate. This makes the glucose more reactive and easier to split as it lowers the **activation energy** of the reaction involved.

1. Oxidative phosphorylation. This occurs when a phosphate ion is added to ADP using energy from electron loss, i.e. oxidation reactions.

2. Photophosphorylation. The energy that powers the proton pump and electron transport chain in chloroplasts comes from light, hence ATP in chloroplasts is synthesised by *photo*phosphorylation.

3. Substrate level phosphorylation. This occurs when phosphate groups are transferred from donor molecules, e.g. phosphate is transferred from glycerate-3-phosphate to ADP in glycolysis of respiration.

Key Term

Activation energy: the energy needed to start a chemical reaction.

quickfire

③ Name a type of cell that contains high numbers of mitochondria.

3.2 Photosynthesis uses light energy to synthesise organic molecules

Overview of photosynthesis

The overall equation for photosynthesis is

$$6CO_2 + 6H_2O \rightarrow C_6H_{12}O_6 + 6O_2$$

Photosynthesis involves two stages:

1. Light-dependent stage where light energy is converted into chemical energy as the photolysis of water releases protons and electrons which produce ATP via **photophosphorylation** and reduce the co-enzyme NADP.

2. Light-independent stage or Calvin cycle where ATP and NADPH from the light-dependent reaction reduce carbon dioxide to produce glucose.

Key Term

Photophosphorylation: the synthesis of ATP from ADP and Pi (inorganic phosphate) using light energy.

quicKfire

④ Name the two products of the light-dependent stage needed by the light-independent stage.

》 Pointer

Photolysis literally means light-split.

separation of molecule by the action of light.

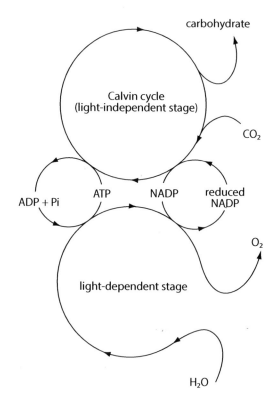

Photosynthesis summary

Structure of the leaf

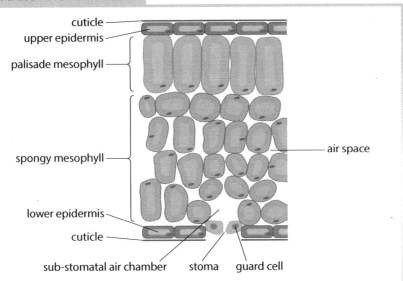

cuticle
upper epidermis
palisade mesophyll
spongy mesophyll
air space
lower epidermis
cuticle
sub-stomatal air chamber   stoma   guard cell

Leaf structure

The leaf is adapted for gas exchange and photosynthesis by having a large surface area allowing the leaf to capture light, and having pores called stomata through which gases diffuse. Air spaces between cells allow for carbon dioxide to diffuse to the photosynthesising cells. The highest concentration of chloroplasts is found in the palisade mesophyll on the leaf's upper surface (up to five times as many as found in spongy mesophyll cells). The palisade cells are arranged vertically, which allows more light to be absorbed by chloroplasts than if they were stacked horizontally, as light only has to pass through the cuticle, epidermal cells and one palisade cell wall.

Structure of chloroplasts

Chloroplasts have a large surface area for the maximum absorption of light. They are able to move within palisade cells to maximise the absorption of light.

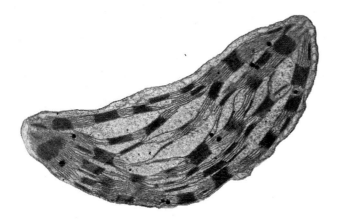

Electron micrograph of a chloroplast

Grade boost

Make sure that you can describe each adaptation of the leaf and explain its significance for photosynthesis. A good way to revise this is to draw a table.

quickfire

⑤ Describe two adaptations that chloroplasts have that maximise the absorption of light.

Grade boost

You should be able to identify the structures present in a chloroplast, and where both stages of photosynthesis take place from either a diagram or an electron micrograph.

Chloroplasts as transducers

The site of photosynthesis was detected by Englemann in 1887. In his experiment, he shone a light through a prism to separate the different wavelengths of light, and exposed this to a suspension of algae with evenly distributed, motile, aerobic bacteria. After a period of time, he noticed that the bacteria congregated around the algae exposed to blue and red wavelengths. This was because this algae photosynthesised more and so produced more oxygen, attracting the motile bacteria.

>> **Pointer**

Blue light has a wavelength of 400–500 nm, red light is 600–700 nm.

Key Term

Transducers: change energy from one form into another.

quickfire

⑥ Why is chlorophyll said to be a transducer?

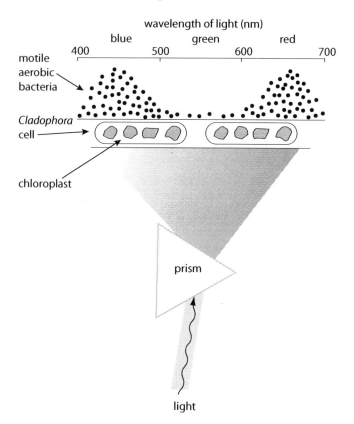

Engelmann's experiment

As **transducers**, chloroplasts can change energy from one form into another, in this case light energy into chemical energy.

Photosynthetic pigments

In flowering plants there are two main types of pigments:

1. Chlorophylls which absorb red and blue-violet regions of the spectrum, e.g. chlorophyll a and chlorophyll b.
2. Carotenoids which absorb light energy from the blue-violet region of the spectrum, e.g. β-carotene and xanthophylls, and act as accessory pigments.

The presence of several pigments allows the plant to absorb a wider range of wavelengths of light than a single pigment.

Absorption and action spectra

The **absorption spectrum** shows how much light energy a particular pigment absorbs at different wavelengths, for example chlorophyll a which absorbs red and blue-violet regions of the spectrum. It does not indicate whether the particular wavelength is used in photosynthesis. An **action spectrum** shows the rate of photosynthesis at different wavelengths, by measuring the mass of carbohydrate synthesised by plants. There is a close correlation between the two, as shown in the graph below.

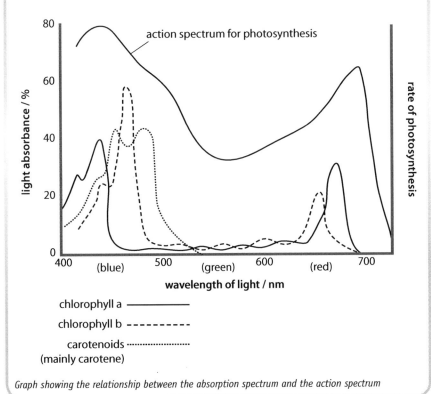

chlorophyll a ————————
chlorophyll b – – – – – – – –
carotenoids ·······················
(mainly carotene)

Graph showing the relationship between the absorption spectrum and the action spectrum

Key Terms

Absorption spectrum: a graph that shows how much light energy is absorbed at different wavelengths.

Action spectrum: a graph that shows the rate of photosynthesis at different wavelengths.

Grade boost

Don't say 'pigments absorb light', you must say 'pigments absorb light energy'.

quickfire

⑦ Using the graph, which pigment has the maximum light absorbance at 450 nm?

quickfire

⑧ What is the difference between absorption spectrum and action spectrum?

Light harvesting

The chlorophylls and accessory pigments are found lying in the thylakoid membranes, grouped into structures called **antenna complexes**. With the aid of special proteins associated with these pigments, light energy (photons) is funnelled towards the reaction centre at the base, containing chlorophyll a. There are two types of reaction centre:

1. Photosystem I (PSI) chlorophyll a, with an absorption peak of 700 nm, also called P700.

2. Photosystem II (PSII) chlorophyll a, with an absorption peak of 680 nm also called P680.

quickfire

⑨ Name the pigment found at the reaction centre of the antenna complex.

quickfire

⑩ What is the role of the accessory pigments?

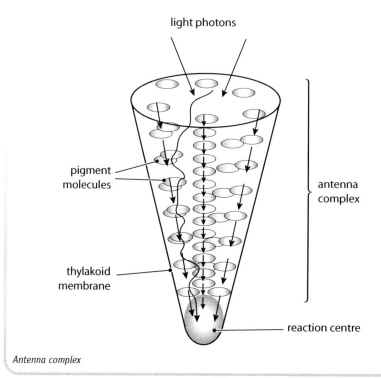

Antenna complex

Identifying the different photosynthetic pigments from chloroplasts

Pigments can be extracted by grinding plant material in a suitable solvent, e.g. propanone, and separated by paper chromatography. By dividing the distance travelled by the pigment by the distance travelled by the solvent front, the R_f value can be calculated.

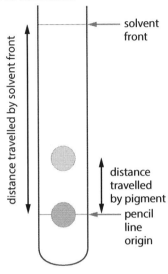

Calculating R_f

The light-dependent stage of photosynthesis

Occurs on the thylakoid membranes. Photophosphorylation occurs via two pathways:

1. Non-cyclic photophosphorylation, which involves both photosystems I and II, generating two ATP molecules and NADPH. Photolysis generates oxygen. The electrons take a linear pathway which is referred to as the 'Z scheme'.

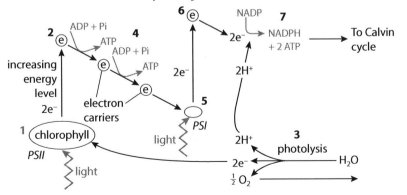

Non-cyclic photophosphorylation

1. Light energy (photons) strikes chlorophyll (PSII) exciting its electrons, boosting them to a higher energy level.

>> **Pointer**
Either reduced NADP, NADPH or NADPH$^+$ / H$^+$ can be used. It may be helpful to use the term NADPH as it allows you to see the movement of protons.

quickfire

⑪ State precisely where cyclic photophosphorylation takes place.

>> **Pointer**

Light does not directly split water, because the photons excite the electrons in chlorophyll boosting them to a higher energy level, which results in chlorophyll becoming oxidised. Oxidised chlorophyll then replaces its lost electrons from water, causing it to split.

quickƒire

⑫ State three differences between cyclic and non-cyclic photophosphorylation.

quickƒire

⑬ What is the source of replacement electrons for photosystem II?

2. Electrons are accepted by an electron carrier in the thylakoid membrane.

3. The oxidised chlorophyll removes electrons from water, producing protons and oxygen (*photolysis*). This occurs in the thylakoid space.

4. As electrons pass from carrier to carrier, electron energy is lost, which pumps protons from the stroma into the thylakoid space. As protons flow back through the stalked particle, ADP is phosphorylated; 2 ATP are made in total.

5. Electrons enter photosystem I where light excites them, boosting them to an even higher energy level.

6. Electrons enter a final electron carrier.

7. Electrons and protons reduce NADP to NADPH which pass to the Calvin cycle with the two ATP made.

2. Cyclic photophosphorylation, which involves only photosystem I, producing 1 ATP molecule only. As photolysis does not occur, no oxygen is released. Electrons take a cyclical pathway.

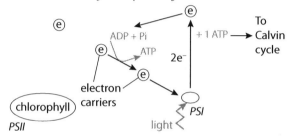

Cyclic photophosphorylation

If there is no NADP available, then the electrons fall back into the electron transport chain (at an intermediate energy level) and generate 1 ATP. This cycle continues until NADP is available. The ATP produced can be used in the Calvin cycle, in the stomatal opening mechanism, or for other cellular processes.

ATP is produced in the chloroplast when protons are pumped across the thylakoid membrane using energy from the electrons and accumulate with protons generated from photolysis of water in the thylakoid space generating an electrochemical (proton) gradient. The H^+ ions diffuse back into the stroma through stalked particles generating ATP. The protons and electrons reduce NADP, which removes H+ ions from the stroma, further contributing to the H+ gradient. The movement of protons is referred to as chemiosmosis.

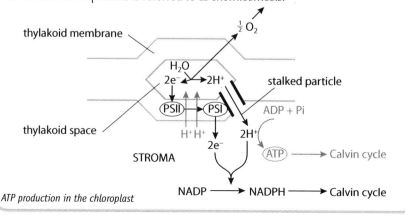

ATP production in the chloroplast

Light-independent stage of photosynthesis (Calvin cycle)

This stage occurs in the stroma. ATP and NADPH from the light-dependent reaction are used to fix carbon from carbon dioxide with the help of the enzyme RuBisCO. The sequence was first worked out by Calvin and his team using a radioactive isotope of carbon (^{14}C) present in hydrogen carbonate. At regular intervals, Calvin removed samples into hot methanol to kill the *Chorella* algae used and to stop all enzyme reactions. He then performed chromatography to identify the products. He exposed his chromatogram to a piece of X-ray film which would detect radiation emitted from ^{14}C used. This identified products containing ^{14}C in the order they were produced: first was hydrogen carbonate ions, then glycerate 3-phosphate (GP), triose phosphate (TP), ribulose bisphosphate (RuBP) and finally glucose.

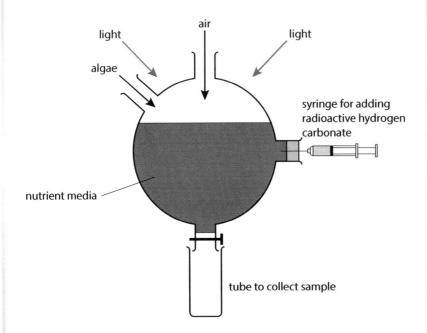

Calvin's lollipop apparatus

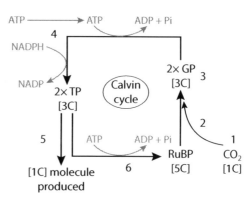

Steps in the Calvin cycle

1. Carbon dioxide diffuses into leaf via stomata, dissolving in the water surrounding palisade mesophyll cells before diffusing into the cells.

2. Carbon dioxide combines with the 5 carbon compound ribulose bisphosphate (RuBP) using the enzyme RuBisCO to form an unstable 6C compound.

3. Unstable 6C compound immediately breaks down into 2 molecules of glycerate 3-phosphate (GP).

4. Using one ATP molecule from the light reaction, GP is reduced to triose phosphate (TP) using hydrogen atoms from NADPH.

5. Triose phosphate molecules combine in pairs to form hexose sugars.

6. Five out of every six triose phosphate molecules produced are used to regenerate RuBP (*via the intermediate ribulose phosphate*) using ATP from the light-dependent reaction to supply energy and phosphate. This allows the cycle to continue.

Product synthesis

Plants must produce all the carbohydrates, fats and proteins they need from the products of the Calvin cycle. Fructose phosphate formed from the two molecules of triose phosphate can be converted to glucose, or combined with glucose to produce sucrose. Sucrose is then translocated in the phloem to the growing regions of the plant. Some α glucose is stored as starch, β glucose forms cellulose in cell walls. Fatty acids can be formed from glycerate 3-phosphate, and glycerol from triose phosphate, the building blocks of triglycerides. Proteins can be formed from glycerate 3-phosphate, but the amino group requires nitrogen from nitrate ions. Other compounds, e.g. chlorophyll, require additional ions e.g. Mg^{2+}, and the middle lamella of cell walls needs Ca^{2+}. A lack of nitrogen results in stunted growth in plants, as plants cannot synthesise proteins due to the lack of nitrogen, whereas a lack of magnesium causes chlorosis, the yellowing of leaves, as chlorophyll cannot be synthesised. This can be shown experimentally by placing plants in soils with different nutrient contents and observing growth.

Grade boost
You should be able to explain the effects of a lack of magnesium or nitrogen on a plant.

Limiting factors in photosynthesis

The rate of photosynthesis is controlled by a number of factors including the concentration of carbon dioxide, light intensity, and temperature. The **limiting factor** is the one which is in shortest supply which controls the rate-limiting step, and therefore an increase in it increases the rate of photosynthesis.

Factor	Graph	Explanation
Carbon dioxide		At low concentrations, carbon dioxide concentration is limiting, but above 0.5%, the rate plateaus, showing that something else must be limiting. Above 1% the stomata close, preventing uptake of carbon dioxide.
Light intensity		As light intensity increases the rate of photosynthesis increases up to about 10,000 lux (SI unit of illuminance) when some other factor becomes limiting. At very high light intensities the rate decreases as chloroplast pigments become bleached. Different plants have evolved to be most efficient at light intensities found in their environment, e.g. sun and shade plants.
Temperature		Temperature increases the kinetic energy of the reactants and enzymes involved in photosynthesis. Unlike other factors, a plateau is not reached as enzymes, e.g. RuBisCO, begin to denature so the rate of photosynthesis decreases above the optimum temperature. This will be higher in species adapted to hot, dry environments.

Grade boost

Although water is a reactant, long before it becomes a limiting factor, the guard cells lose turgidity and the stomata close, reducing the supply of carbon dioxide.
Several factors can be limiting at the same time but the one in shortest supply controls the rate-limiting step.

Key Term

Limiting factor: a factor that limits the rate of a physical process by being in short supply.

Grade boost

The rate-limiting step is the slowest reaction in a sequence, and determines the overall rate of the reaction.

≫ Pointer

The light intensity when the uptake of carbon dioxide is zero, i.e. when the plant is able to use all the carbon dioxide produced in respiration for photosynthesis, is referred to as the light compensation point.
This is different for sun and shade plants.

 Link You covered the effect of temperature on enzyme-controlled reactions in AS. See pages 44–45 in AS Study and Revision Guide.

Grade boost

Remember to perform the experiment three times so a mean can be calculated, which is more reliable than a single result.

Measuring the rate of photosynthesis

Aquatic plants are a good subject to use when investigating how different factors affect photosynthesis. Temperature and carbon dioxide concentrations are more easily controlled than with terrestrial plants, by using a water bath and controlling hydrogen carbonate concentration. It is also easy to collect and accurately measure the oxygen produced in a capillary tube. The volume of the bubble collected is calculated by the formula:

Volume = πr^2 × length of bubble

Where π = 3.14 and r = radius or $\dfrac{\text{diameter}}{2}$

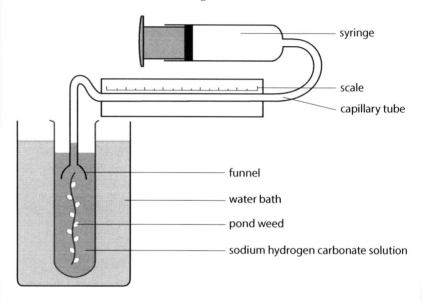

Photosynthometer

Worked example

A photosynthometer with a capillary tube diameter of 0.1 cm was used to measure the volume of oxygen produced by a piece of Canadian pond weed in five minutes at 20°C. The results are shown in the table below:

Temperature / °C	Length of bubble in capillary tube /mm			
	Trial 1	Trial 2	Trial 3	Mean
20	25	20	30	25

Pointer

Remember to work in the same units. In our example, convert 0.1 cm to 1 mm, as the bubble length was also measured in mm, so that the answer is in mm³.

Calculate the mean volume of oxygen produced in mm³.

Volume = πr^2 × length of bubble

diameter = 0.1 cm = 1 mm, so radius = 0.5 mm.
volume = 3.14 × 0.5² × 25
= 3.14 × 0.25 × 25
= 19.6 mm³ (1 dp)

3.2 Complete the following table to show the effects of different factors on the intermediates in the Calvin cycle over a five minute period. Some have been completed for you already.

Factor	Effect on triose phosphate (TP)	Effect on glycerate 3-phosphate (GP)	Effect on ribulose bisphosphate (RuBP)
Light intensity	Decreasing light intensity means less ATP and reduced NADP are made, so less TP is produced (since ATP and reduced NADP are needed to make TP from GP).	Decreasing light intensity means…….	Decreasing light intensity means less ATP and reduced NADP are made, so less RuBP is made because RuBP is still being used up to make GP but RuBP is not being regenerated as less GP is converted into TP, which is needed to make RuBP.
Carbon dioxide concentration	As carbon dioxide increases……	As carbon dioxide increases GP increases, because more CO_2 is fixed, more GP is made.	As carbon dioxide increases……..
Temperature	As temperature increases…….	As temperature increases GP increases. But at high temperatures GP will decrease because the enzyme RuBisCO denatures and less carbon dioxide fixed, so less GP will be made and so less TP is made.	As temperature increases…….

3.3 Respiration releases chemical energy in biological processes

Overview of respiration

The overall equation for **aerobic respiration** is:

$$C_6H_{12}O_6 + 6O_2 \rightarrow 6CO_2 + 6H_2O + 38 \text{ ATP}$$

Aerobic respiration yields a relatively large amount of energy: theoretically up to 38 ATP (although a range of 32–38 ATP is often accepted), and is performed by obligate aerobes. Some microorganisms including bacteria and yeast can respire with or without the presence of oxygen and are termed facultative anaerobes. Bacteria which cannot grow in the presence of oxygen and therefore can only undergo **anaerobic respiration** are termed obligate anaerobes. This will be covered more in the microbiology section.

Aerobic respiration

Aerobic respiration involves four stages:

1. Glycolysis

   - Occurs in the cytoplasm.

   - Does not require oxygen.

   Steps:

   - Glucose is phosphorylated to produce glucose diphosphate. This makes glucose more reactive (by lowering the activation energy of the reactions involved) making it easier to split into triose phosphate.

   - 2 NADs are reduced to NADH when triose phosphate is **dehydrogenated**.

   - 4 ATP are produced by substrate-level phosphorylation, and pyruvate is produced. As 2 ATP are used to phosphorylate glucose the net gain is +2 ATP.

If oxygen is available, the pyruvate moves to the link reaction, and its products move onto the Krebs cycle where more NAD is reduced and some ATP is produced directly.

quickfire

(18) If each NADH yields 3 ATP in the electron transport chain, what is the total ATP yield from glycolysis?

Grade boost

Remember that in total 4 ATP are produced in glycolysis, but as 2 are used to phosphorylate glucose, the net yield is only 2 ATP.

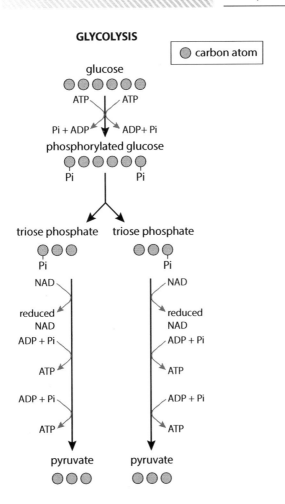

Steps in glycolysis

Key Term

Decarboxylation: the removal of carbon dioxide, performed by decarboxylase enzymes.

2. Link reaction
   - Occurs in the mitochondrial matrix, so pyruvate has to diffuse into the mitochondria.
   - Only occurs in the presence of oxygen.
   - Happens twice *per glucose* molecule (because there are *two* molecules of pyruvate).

   Steps (×2 per glucose):
   - Pyruvate diffuses into the mitochondrial matrix where it is dehydrogenated and the hydrogen released reduces NAD.
   - Pyruvate is **decarboxylated**, producing acetyl.
   - Coenzyme A (CoA) is added to form acetyl CoA which enters the Krebs cycle.

quickғɪre

 Explain the advantage of the highly folded inner membrane (cristae) in mitochondria.

Grade boost

Glucose has to be broken down into pyruvate before it enters the mitochondrion as glucose is too big to diffuse into the mitochondrion, and the mitochondrion does not possess the enzymes needed for glycolysis.

Grade boost

Read any question on Krebs cycle carefully. Is it asking for number of reduced NAD or ATP produced *per molecule* of glucose or *in one* cycle?

Grade boost

You are not required to know the names of all the enzymes just decarboxylases, dehydrogenases and ATP synthetase.

Pointer

Think of the Krebs bicycle i.e. it happens twice!

Grade boost

Without oxygen, the electron transport chain cannot happen as there is no terminal electron acceptor.

3. Krebs cycle

- Occurs in the mitochondrial matrix.
- Only occurs in the presence of oxygen.
- Happens twice *per glucose* molecule (because there are *two* molecules of acetyl CoA).

Steps:

- Acetyl CoA joins to [4C] acid to produce [6C] acid.
- [6C] acid is decarboxylated, releasing 1 molecule of CO_2, and dehydrogenated, reducing 1 NAD molecule.
- The resulting [5C] acid is decarboxylated, releasing 1 molecule of CO_2, and dehydrogenated, reducing 2 NAD and 1 FAD molecules.
- ATP is produced directly by substrate level phosphorylation.
- The resulting [4C] acid combines with acetyl CoA and the cycle repeats.

LINK REACTION and KREBS CYCLE

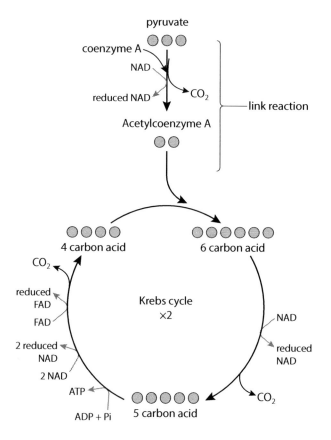

Steps in link reaction and Krebs cycle

4. The electron transport chain
  - Requires oxygen (terminal electron acceptor).
  - Occurs on the inner membrane of mitochondrion (Cristae).

Steps:
  - NADH joins the first proton pump, and is dehydrogenated, releasing the hydrogen atoms which split into protons and electrons.
  - The protons are pumped across the membrane using energy from the high energy electrons as the electrons pass to the next proton pump.
  - As the electrons pass the second proton pump they provide energy to pump a further pair of protons from the matrix to the inter membrane space.
  - The electrons pass the third proton pump; a further two protons are pumped across, which creates a proton gradient.
  - As the electrons pass to the terminal electron acceptor (oxygen), two protons pass back into the matrix through the stalked particle (ATP synthetase) down the proton gradient phosphorylating ADP into ATP.
  - Proton movement here is referred to as chemiosmosis.
  - Water is formed from $2H^+$, $2e^-$ and $\frac{1}{2}O_2$.
  - NADH uses three proton pumps so generates 3 ATP.
  - FADH joins at the second proton pump so only utilises two pumps so only generates 2 ATP.

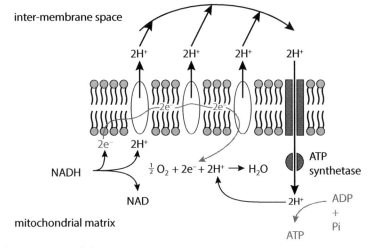

The electron transport chain

Summary of the products of respiration (per molecule of glucose)

Stage	Product				
	ATP	**NADH**	**FADH**	**CO$_2$**	**H$_2$O**
Glycolysis	2	2	0	0	0
Link reaction	0	2	0	2	0
Krebs cycle	2	6	2	4	0
Electron transport chain	34*	0	0	0	6

* From 10 NADH × 3 = 30 ATP and 2 FADH × 2 = 4 ATP

》 Pointer
Cyanide is a non-competitive inhibitor of the final carrier in the electron transport chain and so electrons cannot pass to the terminal electron acceptor. This results in the electrons no longer moving, which prevents the proton pumps from functioning. ATP synthesis soon stops.

3.3 Complete the summary table for the stages in aerobic respiration

Statement	Glycolysis	Link reaction	Krebs cycle	Electron transport chain
Is oxygen needed?				
Is carbon dioxide produced?				
Where does it take place?				
Is FAD reduced?				
Is NADH oxidised?				

Anaerobic respiration

This occurs in many organisms when oxygen is absent. Glycolysis still occurs, but the lack of oxygen prevents link reaction, Krebs cycle and the electron transport chain from occurring. One major consequence of this is that NADH is not oxidised in the electron transport chain, so NAD is not regenerated. As dehydrogenation occurs before production of the final 4 ATPs in glycolysis, the lack of NAD would stop ATP production.

To overcome this, animals can for a short time reduce pyruvate to lactate using hydrogen from NADH, which regenerates NAD, allowing glycolysis to continue. In plants and yeast, the pyruvate is first decarboxylated to ethanal and then reduced to ethanol using the hydrogen from NADH. Lactate and ethanol quickly build up so this cannot be sustained indefinitely. In animals, anaerobic respiration creates an 'oxygen debt' which requires lactate to be oxidised later, releasing further energy. In plants, ethanol cannot be broken down later, so it can accumulate to reach toxic concentrations. In both plants and animals only 2 ATP are produced during anaerobic respiration.

quickfire

 State precisely where the following occur:
a) Glycolysis
b) Link reaction
c) Anaerobic respiration

Grade boost

A build up of lactate in the muscles is toxic and causes cramp. It results in an oxygen debt as it has to be oxidised in the liver later.

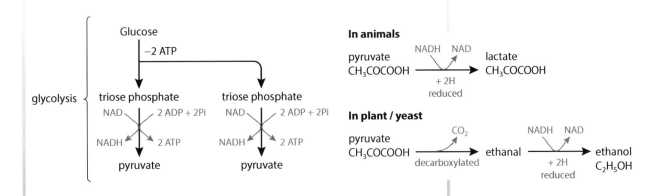

Summary of anaerobic respiration

Efficiency of respiration

One mole of glucose contains 2880 kJ of energy.

The energy liberated from the hydrolysis of ATP = 30.6 kJ per mole.

If we assume a theoretical maximum yield of 38 moles of ATP from a mole of glucose then:

$$\text{Efficiency of aerobic respiration} = \frac{\text{energy from ATP}}{\text{energy in glucose}}$$

$$= \frac{30.6 \times 38}{2880} \times 100 = 40.4\% \ (1 \text{ dp})$$

Grade boost

You can't 'make' energy, only convert if from one form to another, or transfer it.

quickfire

 Calculate the energy efficiency of anaerobic respiration.

Grade boost

Always show your workings in mathematical calculations, and where your answer is not a whole number record to 1 decimal place, unless instructed otherwise.

quicKpire

㉓ Which yields the most energy, 1g of fat or 1g of carbohydrate? Explain your answer.

Alternative respiratory pathways

Lipids can be respired when carbohydrate supplies are low. Lipids are hydrolysed into glycerol which is phosphorylated, and dehydrogenated to form triose phosphate which can enter glycolysis. The fatty acids are split into two carbon molecules that enter Krebs as acetyl CoA. Because fatty acids have large numbers of carbon and hydrogen atoms, respiring them yields more carbon dioxide, water, and ATP (due to more hydrogen being utilised in the electron transport chain).

Proteins can be metabolised when fats and carbohydrates are unavailable, or when diets contain a high proportion of protein. The excess amino acids are deaminated in the liver. The amine group NH_2, is converted to urea in the ornithine cycle in the liver and the urea is then excreted via the kidneys as urine. The carboxyl group that remains can be converted into a number of different Krebs cycle intermediates.

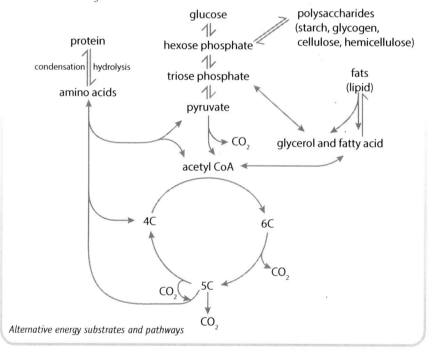

Alternative energy substrates and pathways

Experiments with artificial hydrogen acceptors

A number of artificial hydrogen acceptors, e.g. methylene blue (which turns colourless when reduced) or TTC (which turns red when reduced), can be used with yeast to measure the rate of respiration. The time taken for the colour change to occur can be measured under a number of different independent variables, e.g. temperature or glucose concentration. This method does produce a number of problems: It is difficult to determine the end point, as the time taken for the colour change to occur is subjective, and it produces a reciprocal graph, i.e. a short time represents a high rate of respiration. Using a colorimeter to measure how dark the solution is (% transmission) is a good improvement to determine the end point of the reaction.

3.4 Microbiology

Classification of bacteria

Bacteria are classified according to their shape, cell wall structure, and their metabolic, antigenic and genetic features. Bacteria vary in size from 0.4 μm diameter to over 700 μm, but typically are 7–10 μm long and 2 μm wide e.g. *E. coli*.

Bacteria are usually spherical (coccus), rod shaped (bacillus) or spiral (spirillum), and can be further divided, e.g. *diplo* meaning two. The name often reflects the disease they cause, e.g. *pneumoniae* which causes pneumonia.

Coccus (spheres)

Bacillus (rods)

staphylococi
e.g. *Staphylococcus aureus*
causes food poisoning

e.g. *Salmonella typhi*
causes typhoid fever

diplococci
e.g. *Diplococcus pneumoniae*
causes pneumonia

e.g. *Azotobacter*
is a nitrogen-fixing
bacterium in soil

Different shapes of bacteria

Gram-positive and Gram-negative bacteria

Bacteria are further classified according to the structure of their cell wall, which is determined using the Gram staining technique.

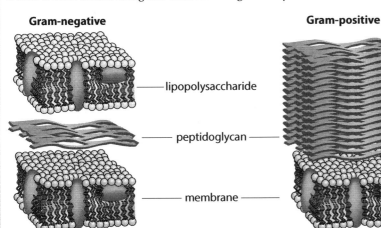

Gram-negative **Gram-positive**

— lipopolysaccharide

— peptidoglycan —

— membrane —

Cell wall of bacteria

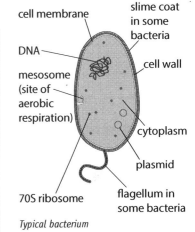

cell membrane
slime coat in some bacteria
DNA
cell wall
mesosome (site of aerobic respiration)
cytoplasm
plasmid
70S ribosome
flagellum in some bacteria

Typical bacterium

>> **Pointer**

Remember the plural terms for the different shapes. Coccus (cocci), bacillus (bacilli) and spirillum (spirilla).

Grade boost

You need to know the Gram stain technique, and be able to explain why Gram-positive bacteria retain the crystal violet stain.

>> **Pointer**

You should be able to identify bacteria from drawings or photographs based upon their shape and colour following Gram stain.

>> **Pointer**

Some books still use the term murein – this is the same as peptidoglycan.

quickfire

㉔ What is the shape of a *Streptococcus* bacterium?

Grade boost

It is important to remember that Gram-negative bacteria *do* contain peptidoglycan in their cell wall, but due to the lipopolysaccharide layer, the crystal violet isn't retained.

Pointer

Remember the P's! Gram-Positive bacteria stain Purple and are susceptible to Penicillin due to the thick layer of Peptidoglycan.

The antibiotic penicillin prevents the cross-links from forming within the peptidoglycan layer, and so weakens the cell wall in newly divided bacteria. Gram-positive bacteria are most affected as they are then subject to osmotic lysis, when water enters the bacterial cell causing the cell to burst.

Gram-positive bacteria	Gram-negative bacteria
Thicker cell wall	Thinner cell wall
Thick layer of peptidoglycan	Thin layer of peptidoglycan
No lipopolysaccharide layer (LPS), so vulnerable to penicillin and lysozyme action	Lipopolysaccharide layer (LPS) protects against penicillin and lysozyme action
Peptidoglycan layer retains crystal violet stain so stains purple	Lipopolysaccharide layer prevents uptake of crystal violet stain, so only stains red once LPS removed and a counter-stain, e.g. safranin used
e.g. *Staphylococcus and Streptococcus*	*Salmonella, and E. coli*

Gram stain

To determine the Gram status of bacteria, they are stained using the Gram stain technique.

1. Transfer a small sample of bacteria to a glass microscope slide using an inoculating loop. Pass the slide through a Bunsen flame a few times to fix the bacteria to the slide (it also kills them).

2. Add a few drops of crystal violet stain and leave for 30 seconds.

3. Rinse excess using water.

4. Add Gram's iodine for 1 minute to fix stain.

5. Bacteria which stain purple are Gram-positive.

To stain remaining bacteria:

1. Wash with alcohol for 30 seconds to dissolve lipids in lipopolysaccharide layer, and expose inner peptidoglycan layer.

2. Re-stain using another stain, e.g. safranin which stains unstained bacteria red.

quickfire

㉕ Explain the purple colour seen when Gram staining *Staphylococcus* bacteria.

quickfire

㉖ What layer is only present in Gram-negative bacteria?

Conditions necessary for bacterial growth

A growth medium such as nutrient agar has the following:

- Nutrients – a source of carbon for respiration, e.g. glucose, nitrogen for synthesis of nucleotides and proteins, and vitamins and mineral salts.

- Water.

- Suitable temperature – 25–45°C for most bacteria; 37°C is optimum for mammalian pathogens. Some can survive at 90°C (these are called thermophiles), e.g. *Thermus aquaticus* which evolved in hot springs.

- Suitable pH – optimum is slightly alkaline (pH 7.4) for most bacteria. Some can survive acidic conditions e.g. *Helicobacter pylori* in stomach (pH 1–2).

- Oxygen may or not be required depending upon the mode of respiration.

- If a microbe needs oxygen for metabolism it is termed an **obligate aerobe**.

- Those which grow better in the presence of oxygen BUT can grow without it are **facultative anaerobes**.

- Those which CANNOT grow in the presence of oxygen are called **obligate anaerobes**, e.g. *Clostridium botulinum* bacteria, which produce the botulinum toxin.

Key Terms

Obligate aerobes: microbes that require oxygen for growth.

Facultative anaerobes: microbes that grow better with oxygen but can grow without it.

Obligate anaerobes: microbes that cannot survive in the presence of oxygen.

quickfire

㉗ Why is a source of nitrogen added?

Aseptic technique

When culturing bacteria it is important to use **aseptic technique** to ensure that only the desired bacterium is grown, and that you don't contaminate yourself or the environment. Equipment and media used must be sterilised by using:

- Heat at 121°C for 15 minutes in an autoclave or pressure cooker, or by passing the equipment through a Bunsen flame for 2–3 seconds until it glows red e.g. an inoculating loop. This works for inanimate objects (non-living).

- Irradiation works well for heat-labile plastics.

- Benches cannot be sterilised but can be disinfected, e.g. with 3% Lysol, which reduces numbers of microbes, BUT NOT fungal spores.

- Living tissues cannot be safely sterilised without killing them so antiseptics are used which kill or inhibit microbes on the outside of living tissues only.

It is important to grow bacteria at 25°C rather than 37°C so that **pathogenic** microorganisms aren't grown. Petri dish lids should be secured with tape but not completely so oxygen can still get in. All material should be safely disposed of afterwards by sterilising in an autoclave.

>> *Pointer*

Heat-labile plastics melt at the temperatures needed to sterilise them, so radiation has to be used instead.

Grade boost

Sterilising kills all microorganisms including spores. Disinfection reduces the number of microbes.

Key Terms

Aseptic technique or **sterile technique**: good laboratory practice that maintains sterile conditions and prevents contamination.

Pathogen: a disease-causing microorganism.

>> *Pointer*

Colony Forming Unit (CFU): We assume that each colony grew from a single bacterium.

◀Link▶ Population growth is also covered in Section 3.5.

Measuring bacterial growth

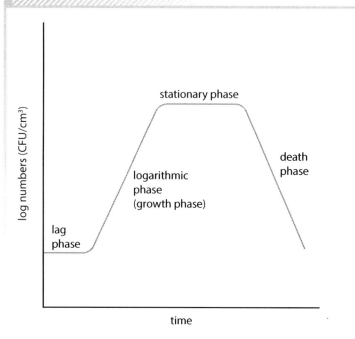

Hypothetical growth curve

During the Lag phase, population number increases very slowly because time is needed for enzyme synthesis.

During the Log/Exponential/Growth phase, there are plenty of nutrients and few toxic by-products so there are no limiting factors. This allows rapid reproduction.

During the stationary phase, cells are reproducing but population is relatively constant fluctuating around the carrying capacity, due to cell production equalling cell death. The population has reached its carrying capacity because reduced resources (e.g. nutrients/space/toxic waste products) are now limiting factors.

During the death phase more cells are dying than are being produced so the population decreases. Death of cells is due to lack of nutrients, lack of oxygen or increased toxicity of the medium.

Growth can be measured in two main ways:

1. Directly where the total number of cells is calculated.

2. Indirectly by measuring the turbidity (cloudiness) of a culture.

Direct counts can be either 'viable counts' where only living cells are counted or 'total counts' where both living and dead cells are counted, by using a haemoctyometer, originally developed to count blood cells.

>> *Pointer*

You will not be required to detail the use of a haemocytometer or colorimeter in the exam, but you should know the ways in which growth can be measured.

Viable cell counts

This counts the number of living cells and is particularly useful in medical and food hygiene applications. Even in small cultures, the total viable cell count can exceed several million per cm^3, so a serial dilution must be performed first. This is often done in tenfold steps, i.e. 1 in 10, but higher dilutions can be performed, e.g. 1 in 100. For a 1 in to 10 dilution, 1 cm^3 is added to 9 cm^3 of sterile medium and mixed, and repeated until a range of dilutions is obtained. The next step is to plate out each dilution and incubate the plates at 25°C for 24–48 hours to allow bacteria to grow. The plates are examined and a plate chosen to count: the best one is a plate containing between 20 and 100 colonies (more than this is difficult to count, and fewer than 10 has an increased error due to the small numbers involved). The viable cell count is calculated by multiplying the dilution factor by the number of colonies. This technique has to assume that each colony originated from a single bacterium that divided asexually.

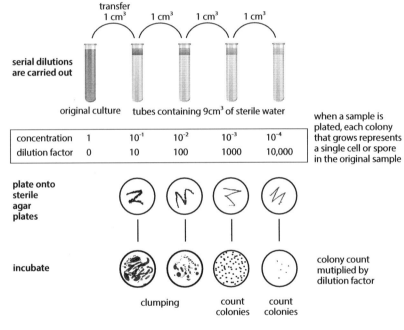

concentration	1	10^{-1}	10^{-2}	10^{-3}	10^{-4}
dilution factor	0	10	100	1000	10,000

when a sample is plated, each colony that grows represents a single cell or spore in the original sample

colony count mutiplied by dilution factor

Serial dilution and plating

Worked example

To work out the total number of bacteria, multiply the number of colonies on the plate by the dilution factor, e.g. if a plate contained 59 colonies from a 100 000-fold dilution (10^{-5}), the number of bacteria present per cm^3 of the original culture is 59 × 100 000 = 5 900 000 or 5.9 million. It is best to represent this as $5.9 × 10^6$.

Grade boost

For dilution plate calculations always show your working and read the question carefully: Is it asking for the total number of bacteria per cm^3 or in the total culture?

Grade boost

A common mistake occurs when only 0.1 cm^3 of sample is spread. This represents a further tenfold dilution and so you should multiply the number of colonies by the dilution factor and then by 10.

quickfire

㉘ What is the difference between a viable and total cell count?

quickfire

㉙ 1 cm^3 of a 25 cm^3 culture was diluted ten times by diluting 1 cm^3 into 9 cm^3 of sterile medium; 0.1 cm^3 of each dilution was plated out. The 10^{-2} dilution plate had over 200 colonies, the 10^{-3} dilution plate had 20 and the 10^{-4} dilution plate had 2 colonies present.

a) What is the main assumption made when performing a dilution plate count?

b) Calculate the total number of bacteria in the original culture.

3.5 Population size and ecosystems

Key Terms

Population: the total number of organisms of a single species interbreeding within a **habitat**.

Habitat: the physical place where an organism lives.

Birth rate: the number of new individuals derived from reproduction per unit time.

Immigration: the movement of individuals into a population.

≫ Pointer

Birth rate and immigration increase population size, whereas death rate and emigration reduce population size.

quickfire

㉚ Complete a table to compare the factors influencing each of the four phases in population growth for
a) a bacterial population,
b) a mammalian population.

The number of individuals in a **population** changes over time. It can be represented by the equation:

Population size = **birth rate** + **immigration** − death rate + emigration

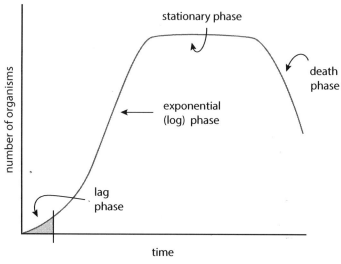

Graph showing changes in population growth

During the lag phase there is a period of slow growth. In sexually reproducing organisms e.g. rabbits, this represents the time taken to reach sexual maturity and gestating young.

During the log phase (growth phase), numbers increase logarithmically as there are no factors limiting growth. This cannot be maintained indefinitely as environmental resistance reduces growth from the availability of food and space, and other **biotic factors** such as predation, competition, parasitism and disease, and **abiotic factors** such as soil pH, light intensity and temperature which also reduce population growth.

During the stationary phase birth and death rates are equal and the population has reached its maximum size, or **carrying capacity**. Numbers will fluctuate around this in response to environmental changes. This is often due to predator-prey relationships, where **negative feedback** regulates, i.e. number of prey decrease so there is less food so number of predators decrease, which reduces predation, so prey number increase, and so on. These fluctuations exist over months, even years as population responses are slow.

During the death phase, factors that have reduced population growth become more significant and the population size decreases. Death exceeds births.

Calculating population increase from a graph

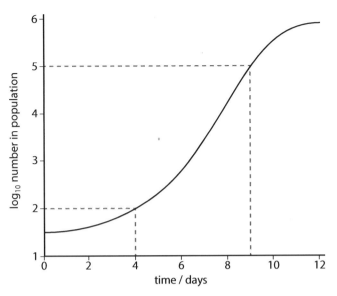

Bacterial growth curve

$$\text{Rate of growth per day} = \frac{\text{antilog}_{10}5 - \text{antilog}_{10}2}{5}$$

$$= \frac{100\,000 - 100}{5}$$

$$= 19\,980 \text{ per day.}$$

Density-dependent and density-independent factors

Some factors have an increased effect on larger population sizes, i.e. a denser population, and are called density-dependent factors, and are biotic factors, e.g. predation and disease. In larger populations, disease is more easily spread, and a predator can find more prey more easily. Density-independent factors are abiotic, e.g. light intensity or temperature, and therefore their effect is the same regardless of the population density. Fire is another example: it will kill all life in its path, whether it is one tree or hundreds!

Grade boost

When writing about population growth it is important to consider the organism. Bacteria are not born, mammals are. The death phase in bacteria occurs largely due to build up of toxic waste products; the same cannot be said about a population of rabbits!

Key Terms

Biotic factor: a living factor, e.g. a predator or pathogen, which can influence the population.

Abiotic factor: a non-living factor, e.g. oxygen availability or air temperature, which can influence the population.

Carrying capacity: the maximum number around which a population fluctuates in a given environment.

Negative feedback: occurs in an equilibrium where the corrective mechanism is in the opposite direction to the direction of change, e.g. if population numbers increase, negative feedback results in a decrease and vice versa.

≫ Pointer

A log scale is used to show very large numbers which a linear scale would be unable to do. With $\log_{10}$ the scale increases by a factor of ten times each time.

≫ Pointer

To find the antilog of 6.7 on your calculator, press SHIFT log, press 6.7, press = for the answer (5.0×10^6) (1 dp).

Measuring abundance and distribution in a population

The abundance of a species is a measure of how many individuals exist in a habitat. It can be assessed by a range of techniques. Physical features like soil type, pH and temperature will influence the range of organisms that can live there. Where conditions are optimal, e.g. warm, good rainfall, high sunlight intensity, then many plants will be found supporting many other animals.

Different sampling techniques

To estimate the number of individuals of each species in a given area, a number of practical techniques can be used. Sampling should be at random to eliminate sampling bias.

Population	Technique	Method
Terrestrial animals	Mark-release-recapture (Lincoln Index)	Animals are captured and marked (it is important that they are not harmed or made more visible to predators) and then released. Once animals have had chance to reintegrate with the population, e.g. 24hrs, the traps are reset. The total population size can be estimated using the number of individuals captured in sample 2, and the number in that sample that are marked (i.e. caught before). $$\text{Pop size} = \frac{\text{no. in sample 1} \times \text{no. in sample 2}}{\text{no. marked in sample}}$$ Have to assume that no births/deaths/immigration/emigration, have occurred during the time between collecting both samples.
Freshwater invertebrates	Use kick-sampling and use Simpson's Index	Collect and identify invertebrates from a given area using a quadrat and a net. Kick or rake the area e.g. 0.5 m² for a set period, e.g. 30 seconds, and collect invertebrates in a net downstream. Release invertebrates carefully. Use Simpson's Index to calculate diversity.
Plants	Quadrats and transects	Estimate percentage area cover of different plants using a quadrat divided into 100 sections. Measure plant density by counting number of plants in a quadrat, e.g. 1m². A transect is a tape measure that is used to measure intervals along an environmental gradient, e.g. distance from a woodland, along which quadrats can be placed.

Ecosystems

An **ecosystem** represents the total number of different organisms of all species present in a habitat in which energy and matter are transferred in complex interactions between the environment and organisms. Examples include tropical rainforest, temperate deciduous forest, tundra and desert. The abiotic and biotic features vary from ecosystem to ecosystem.

Food chains

A food chain represents the energy flow through an ecosystem. The ultimate source of energy for a food chain is sunlight which is converted into chemical energy by **producers** via photosynthesis. Most of the energy available at each **trophic level** is released in respiration and incorporated into other molecules or into electrochemical gradients. This means that often less than 10% is incorporated into **biomass** and is available to the next trophic level, which ultimately limits the length of food chains.

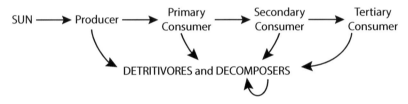

Typical food chain

Primary consumers are herbivores and feed on producers. Secondary and tertiary consumers include carnivores and feed on the trophic level below. Food chains are often more complicated than this as some tertiary consumers feed at more than one trophic level, so these more complex energy flow diagrams are represented as food webs. Decomposition involves detritivores, e.g. earthworms and woodlice, which feed on detritus (the remnants of dead organisms and fallen leaves) and decomposers, e.g. bacteria and fungi, that feed via external digestion (saprotrophism) completing the process started by the detritivores. They therefore feed on all trophic levels. When decomposers die they are fed on by by other decomposers.

Efficiency of photosynthesis

The majority of light falling on a plant (60%) may not be absorbed by the pigments within the chloroplasts because it is:

- The wrong wavelength.
- Reflected by the leaf surface.
- Transmitted through the leaf without striking a chlorophyll molecule.

Less than 1% of energy from sunlight is fixed, leaving around 0.5% or less being available to the next trophic level as biomass. For many wild plants this is much lower, just 0.2%. The photosynthetic efficiency can be calculated by:

$$\text{Efficiency} = \frac{\text{Quantity of light energy fixed by plant}}{\text{Quantity of light energy falling on plant}} \times 100$$

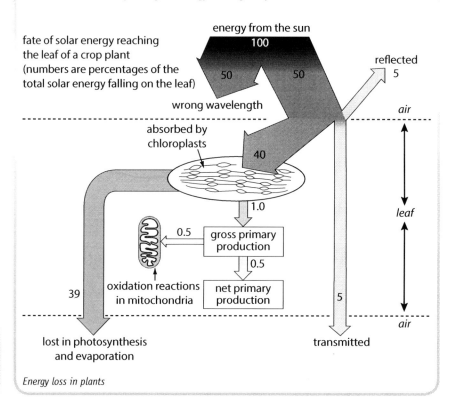

Energy loss in plants

>> **Pointer**

Less than 0.5% of light falling on a plant is converted into biomass.

quickfire

㉞ Explain why most of the sun's energy is not transferred to a plant.

Primary productivity

GPP or **gross primary productivity** is the energy fixed by green plants (photosynthesis) in a given area over a given time, i.e. kJ m^{-2} year^{-1}. Not all the energy is available for next trophic level, so we use NPP or **net primary productivity** which is a measure of what is actually available for animals to eat (the plant's biomass). This can be represented as:

$$\textbf{NPP = GPP} - \textbf{R} \quad \text{where R = Respiration}$$

Some of the biomass is used to form inedible material, e.g. bark, or is biomass in roots which is out of reach of primary consumers, therefore the true value available is even lower. NPP varies according to the ecosystem: tropical rainforests have a very high NPP due to plentiful rainfall, high light intensities and warm temperatures. Tundra has a much lower NPP due to the environment being cold, with much lower light intensities.

Energy flow through food chains

The transfer of energy in the form of biomass from one trophic level to the next is relatively low at around 10% or less. In a primary producer, up to 60% is lost typically in excretion and egestion (urine and faeces), and 30% is lost as heat in respiration. The proportion of chemical energy of food which consumers convert into biomass is referred to as the secondary production. In carnivores this is much higher at around 20%, due to the fact that they can digest their protein-rich diet more efficiently, and is important because the energy available at the end of a food chain is very small. Whilst up to 60% of energy is lost in faeces and urine from a primary consumer, in a top carnivore it is just 20%. This is largely due to the cellulose-rich diet in herbivores, which despite several adaptations (cellulose-digesting bacteria, chewing the cud, four chambers) see AS SRG p112–114, much energy is not obtained.

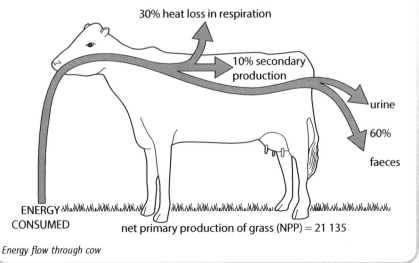

Energy flow through cow

Key Terms

Gross primary productivity: the rate of production of chemical energy in organic molecules by photosynthesis in a given area, in a given time, measured in kJ m^{-1} y^{-1}.

Net primary productivity: the energy in the plant's biomass which is available to primary consumers, measured in kJ m^{-1} y^{-1}.

>> **Pointer**

GPP in many systems is as low as 0.2%, with NPP at 0.1%.

Grade boost

Only part of the NPP in an ecosystem is transferred to the primary consumers due to low conversion efficiency.

Calculating energy efficiency

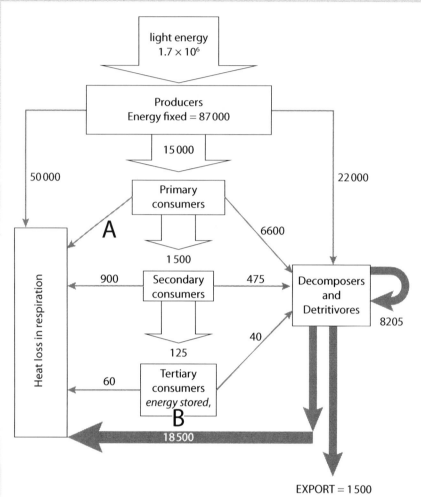

All values in kJ m^{-2} yr^{-1}

Energy flow through an ecosystem

③⑤ How much energy is lost in respiration by the primary consumers (labelled A)?

③⑥ How much energy is stored by the tertiary consumers (labelled B)?

③⑦ Using your answer to Quickfire 36, calculate how efficient the carnivores are at assimilating energy into their bodies.

To calculate the % efficiency of energy transfer use the equation:

$$\% \text{ efficiency} = \frac{\text{Energy fixed as biomass}}{\text{Energy available to next trophic level}} \times 100$$

In the diagram above, to calculate the % efficiency for the secondary consumers:

$$\% = \frac{125}{1500} \times 100$$

$$= 8.3\% \text{ (1 dp)}$$

Ecological pyramids

A pyramid of numbers is relatively easy to construct and shows the energy flow through a food chain: as energy is lost at each stage, fewer individuals can be supported. However, it doesn't take into account the size of organisms, for example a few wheat plants can support large numbers of greenfly, and so the pyramid inverts at this trophic level. Also, when dealing with very large numbers, e.g. with aphids, it is difficult to draw the bars to scale.

A pyramid of biomass is more accurate, but is difficult to measure! These pyramids may also be inverted, which can happen when organisms have a rapid life cycle and so numbers are replenished very quickly, e.g. phytoplankton, where total biomass over a year is considerably higher than at any given time. These pyramids are also difficult to calculate (how do you measure the mass of tree roots?).

The most accurate way of representing feeding relationships is to use pyramids of energy. They show more clearly the energy lost at each level, but are difficult to calculate. No pyramid, however, can show that some organisms operate at several trophic levels.

>> **Pointer**

Only a pyramid of energy shows true energy flow through an ecosystem.

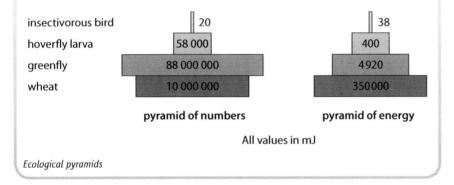

Ecological pyramids

	pyramid of numbers	pyramid of energy
insectivorous bird	20	38
hoverfly larva	58 000	400
greenfly	88 000 000	4 920
wheat	10 000 000	350 000

All values in mJ

Community and succession

Ecosystems are dynamic and subject to change over time. The change in the composition of a community over time is referred to as **succession** and occurs over tens to thousands of years depending upon the starting point.

Each stage in succession is known as a seral stage where different communities dominate as they outcompete pre-existing species as conditions are more favourable for them.

Primary succession occurs when organisms invade spaces that did not previously support life, e.g. bare rocks, sand dunes, volcanic flows.

Key Term

Succession: the progressive changes in the structure and species composition in a community over time.

Weathering creates small cracks in the rocks and small particles.	Mosses and lichens begin to colonise. Organic matter builds up slowly.	Legumes begin to grow as they are able to fix atmospheric nitrogen to supplement the poor nutrient soil. As they die, soil becomes enriched.	Grasses and ferns start to grow, sheltering the soil from the elements. Soil, and its moisture content increases.	Large shrubs and small trees colonise. Leaf litter greatly increases fertility and humus content of the soil. Habitats created for nesting birds and soil invertebrates so diversity increases.	Climax woodland is reached. This is usually oak, beech, hazel or lime species but is largely deciduous in Southern UK. Ground flora includes bracken, shrubs and bluebells.

Primary succession from bare rock

What changes as succession progresses?

- Soil depth increases
- Nutrient content increases
- Humus content increases therefore water content increases
- Species diversity increases
- Stability of community increases.

Secondary succession is the reintroduction of organisms into a habitat previously occupied by plants and animals.

1. Soil is present.
2. Disrupted from succession by event such as fire/flooding/cultivation.

A deflected or disclimax arises when the **climax community** is not reached due to practices such as monoculture, or grazing, i.e. human intervention. Heather moors are managed to increase grouse numbers by routinely burning large areas every twelve years to remove old growth and encourage new **pioneer species** which provide food for grouse.

Comparison of primary and secondary succession

Primary succession	Secondary succession
Surface is bare	Soil present
Pioneer organisms are lichens and mosses	Pioneer organisms are typically small weedy plants
Takes a long time to reach climax community because soil must be created through physical and biotic interactions	Quicker to reach climax community

Key Terms

Climax community: a stable community that undergoes no further change.

Pioneer species: the first species to colonise a new area in an ecological succession, e.g. Mosses and lichens in primary succession.

quickfire

㊳ Match the following terms to their correct statements:

1 Seral stage.
2 Climax community.
3 Pioneer community.
4 Secondary succession.

A The recolonising of land following a fire.
B First organisms to colonise the area.
C Stable community in which there is no further change.
D A stage in succession.

Factors affecting succession

The immigration of spores and seeds into an area are important in the recolonisation of an area. As new species are introduced, competition exists for resources at all the seral stages because, for example, legumes can outcompete mosses as the soil content increases. Competition exists between:

1. Different species (interspecific competition) where each may occupy a different **niche**.

2. Individuals of the same species (intraspecific competition) which is density dependent, i.e. competition increases with population size.

A good example of interspecific competition is shown between two species of *Paramecium*, *P. aurelia* and *P. caudatum*. A Russian scientist, Gause, cultured both species separately and found they had similar growth curves; however, when cultured together, the smaller, faster-growing *P. aurelia* outcompeted the larger, slower-growing *P. caudatum*. Gause formulated the 'Competitive exclusive principle' that states that when two species occupy the same habitat, one will out-compete the other – in other words two species cannot occupy the same niche.

Some species exist in a relationship that is beneficial to both which is called **mutualism**. An example is the nitrogen-fixing bacterium *Rhizobium*, which inhabits the root nodules of leguminous plants: it receives a source of carbon from the plant and in return supplies the plant with nitrogen compounds from which it can synthesise nucleotides and proteins.

Some species are able to obtain benefit from one species whist the other is unaffected, which is referred to as **commensalism**, e.g. the small fish that attach themselves to larger fish for locomotion and food scraps.

quickfire

(39) What is the difference between mutualism and commensalism?

Recycling nutrients

Microorganisms are important in the recycling of a number of nutrients, e.g. carbon and nitrogen, which cycle between the biotic and abiotic components of an environment.

The carbon cycle

Carbon is the building block of life: it is a major component of carbohydrates, fats and proteins, and found in many other molecules. It is absorbed from the atmosphere during photosynthesis and returned during respiration. Carbon dioxide is added to the air during combustion of fossil fuels, but decreasing amounts are being removed by photosynthesis as large areas of forest are being removed and the land used for other purposes (**deforestation**), resulting in unprecedented rises in atmospheric carbon dioxide levels.

It is important to remember that detritivores and decomposers feed on every trophic level, and as they respire, carbon dioxide is returned to the atmosphere.

Carbon dioxide is dissolved in aquatic ecosystems as HCO_3^- ions, and undergoes similar processes as in the atmosphere, except that it forms calcium carbonate in mollusc shells and arthropod skeletons. When these organisms die, and their shells settle on the ocean bed, compression over millions of years forms chalk, limestone and marble, from these carbonates which act as a long-term store (or sink) for carbon. Erosion of these rocks can return carbon dioxide back to the atmosphere.

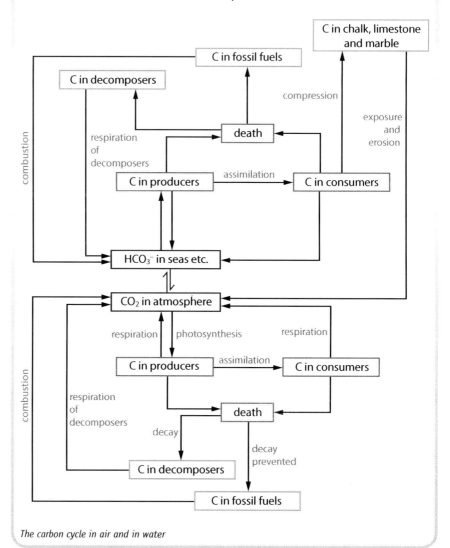

The carbon cycle in air and in water

» Pointer

Between 1900 and 2015 the atmospheric concentration of carbon dioxide has increased from 290 ppm to nearly 400 ppm at an increase of nearly 38%.

quickfire

⑩ Explain why planting trees (afforestation) would reduce atmospheric carbon dioxide.

Human impact on the carbon cycle

Deforestation has resulted in up to 40% reduction in tree cover since the industrial revolution, significantly reducing the volume of carbon dioxide removed from the atmosphere through photosynthesis.

The burning of fossil fuels accounts for the majority of the increases in atmospheric carbon dioxide levels seen in the past 150 years. Carbon dioxide is a greenhouse gas which absorbs infra-red radiation and re-radiates it back towards the Earth's surface rather than allowing it to escape into space. We need a natural greenhouse effect, or temperatures would fluctuate too widely to support life, but increases in carbon dioxide emissions over recent decades are leading to an enhanced greenhouse effect. Other gases, e.g. methane, nitrous oxide, ozone, water vapour and CFCs, act as strong greenhouse gases and their levels have been increasing since the 1900s. Some of these gases are more warming molecule for molecule than carbon dioxide, e.g. methane is 25 times and nitrous oxide is nearly 300 times more warming, but due to the very high concentrations of carbon dioxide involved, carbon dioxide is still the major concern. A major source of methane comes from rice production in paddy fields and from cattle and other livestock.

The following table summarises the activities and their roles in carbon dioxide levels.

Activity / process	Result on global CO_2 levels
Photosynthesis	Decreases
Respiration	Increases
Combustion	Increases
Deforestation	Increases

Rising carbon dioxide levels lead to:

1. **Global warming** is a result of the enhanced greenhouse effect, and could be as high as 5°C by the end of the century, but even conservative estimates place this at 2°C. As global temperatures rise, polar ice will melt, resulting in sea levels rising and coastal flooding, and higher temperatures increase the incidence of forest fires, and lead to **desertification**.

2. Climate change is a consequence of global warming as changes to regional climate patterns, average temperature, wind patterns and rainfall become more noticeable, and extreme weather conditions, e.g. drought and hurricanes, become more frequent. As the climate changes, plants and animals may not be able to adapt or migrate and many will become extinct. There is a risk from both global warming and the associated change in climate that many areas will suffer crop yield reductions and failed harvests, unless farming practices are changed, e.g. through use of drought-tolerant crops. Acidification of oceans due to increasing dissolved carbon dioxide will affect many aquatic organisms: fish gills produce mucus in response to increasing acidity which reduces gas exchange, and crustaceans lose calcium carbonate from their exoskeletons as it is soluble in acid.

Key Terms

Global warming: the increase of average global temperatures in excess of the greenhouse effect caused by the atmosphere's historical concentration of carbon dioxide.

Desertification: the process by which fertile land becomes desert as it loses water, vegetation and wildlife.

Carbon footprint

The **carbon footprint** represents the equivalent amount of carbon dioxide produced in one year by an individual or activity. Agriculture, whilst removing carbon dioxide via photosynthesis, incurs a carbon footprint as energy is needed to produce insecticides and fertilisers, fossil fuels power farming machinery, and produce needs to be transported to market. To meet consumer needs for out of season produce, some crops travel thousands of miles, e.g. blueberries from Chile, roses from Kenya. By firstly reducing use of products, then reusing products more, and finally recycling (the three R's) we can all reduce energy consumption and hence our carbon footprints.

The nitrogen cycle

The nitrogen cycle involves the flow of nitrogen atoms between the atmosphere and inorganic and organic nitrogen compounds in the soil or water. Nitrogen is needed by plants to synthesise nucleic acids, and proteins. Because nitrogen gas is inert, it must be absorbed by plants in some other form, usually nitrate ions in solution via their roots. Animals obtain a source of nitrogen by digesting plant and animal proteins.

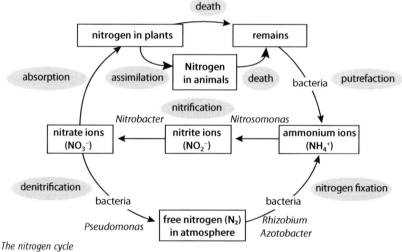

The nitrogen cycle

The four main processes in the nitrogen cycle are:

- Ammonification (putrefaction): bacteria and fungi digest extracellularly dead organisms, faeces and urine. Proteases hydrolyse proteins to amino acids and deaminases reduce the amino groups to ammonium ions (NH_4^+).
- **Nitrification:** the addition of nitrates to the soil by the conversion of ammonium ions to nitrites by *Nitrosomonas* bacteria and then the conversion of nitrites to nitrates by *Nitrobacter* bacteria. The first reaction involves the loss of hydrogen atoms and both result in the gain of oxygen, meaning both reactions are oxidation and require aerobic conditions.
- Denitrification: the loss of nitrate from the soil by anaerobic bacteria *Pseudomonas*, to atmospheric nitrogen.

- **Nitrogen fixation:** the reduction of atmospheric nitrogen molecules to ammonium ions. This is accomplished by two genera of bacteria:
- *Azotobacter* free living in the soil accounts for most nitrogen fixation.
- *Rhizobium* is a mutualistic bacterium found in the root nodules of leguminous plants, e.g. pea and clover. Nitrogen gas diffuses into the nodules where nitrogenase enzyme produced by the bacteria reduces nitrogen (N_2) to ammonium ions (NH_4^+) in an anaerobic process. The ammonium ions are converted to organic acids and amino acids for the bacteria, and some enter the phloem for use by the plant.

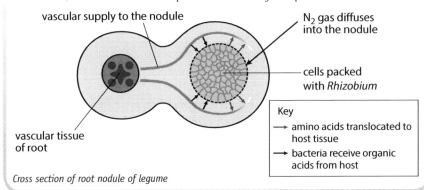

Cross section of root nodule of legume

Human impact on the nitrogen cycle

To maximise yields, farmers use pesticides to reduce damage to crops, and use nitrogen-based fertilisers, e.g. ammonium nitrate, to improve plant growth. Farmers regularly plough and drain soils to increase soil aeration, which favours the aerobic processes of nitrogen fixation and nitrification, whilst inhibiting denitrification, which is anaerobic. This also allows air to reach the roots of plants, as nitrates and other minerals are taken up by active transport, which requires ATP (from aerobic respiration). Manure and slurry are also added to soils to improve soil structure and nitrogen content.

A major consequence of using water-soluble nitrate fertilisers is that they can be washed off (leached) into water courses, e.g. rivers and streams, which increases the ion content – this is called **eutrophication**. Phosphates used in washing powders entering the water course further compound the problem.

The increase in nitrates and phosphates leads to an algal bloom which blocks out light to aquatic plants. As plants and algae die, they form detritus, which is decomposed aerobically, using up dissolved oxygen in the water, which causes other aerobic organisms, e.g. fish, to die, further adding to the detritus. Eventually all the oxygen is used up and only anaerobic bacteria survive, releasing ammonia, methane and hydrogen sulphide into the water creating an environment toxic to most aerobic organisms.

To reduce fertiliser run-off farmers can:

1. Apply fertilisers when plants are actively growing. This increases their uptake reducing build up in the soil.
2. Not apply fertilisers within 10 metres of watercourses.
3. Dig drainage ditches to collect any run off.

Grade boost
You should be able to name the genera of the bacteria involved in the four processes found in the nitrogen cycle.

Grade boost
Remember that nitrogen-fixing bacteria reduce atmospheric nitrogen (N_2) to ammonium ions (NH_4^+) in an anaerobic process.

quicKfire
㊶ Explain the role of nitrogenase enzymes in nitrogen fixation.

quicKfire
㊷ Explain why it is important to maintain aerobic conditions in the soil.

Key Term
Eutrophication: the artificial enrichment of aquatic habitats by excess nutrients, often caused by run-off fertilisers.

Pointer
Biological oxygen demand (BOD) is the amount of dissolved oxygen needed by organisms to break down organic material present in a given water sample at certain temperature over a specific time period. When this is high, microorganisms are respiring aerobically depleting the dissolved oxygen in the water.

3.6 Human impact on the environment

Why species are at risk

Human activities such as deforestation, agriculture, overfishing, pollution, forestry, mining and urban expansion have all resulted in widespread habitat destruction, resulting in a biodiversity crisis as species compete for resources. Many species are becoming endangered and some have become **extinct** because of habitat destruction or through hunting and collecting, e.g. ivory.

Species extinction can be caused by:

(i) Changes in climate which reduce vegetation and decreases atmospheric oxygen levels.

(ii) Human activity, e.g. the dodo bird was indigenous to Madagascar and Mauritius, but habitat destruction and hunting by sailors resulted in the bird's extinction by 1662.

Endangered species

An **endangered species** is one that is at risk of extinction in either all or a significant part of its range (where it is found). Species are classified according to their vulnerability to extinction: critically endangered, endangered, and vulnerable. Threatened species today include mountain gorillas, giant pandas and polar bears.

Reasons why species are at risk:

- Natural selection: Requires mutations in the gene pool to confer a selective advantage. Where species cannot adapt to changes in their habitats fast enough due to insufficient mutations they can become threatened or extinct.

- Habitat destruction: E.g. deforestation and hedgerow removal. Hedgerows contain many different plant species supporting a wide variety of animal life, but their removal to allow for larger fields to accommodate agricultural machinery and the subsequent loss of wildlife corridors has reduced species numbers and affected food chains.

- Pollution: E.g. PCBs (polychlorinated biphenyls) were manufactured as coolants and have since been banned due to their toxicity and carcinogenic nature. They are still found in the environment close to manufacturing sites.

  - Oil is shipped worldwide to meet our energy needs, but accidents at sea have led to the discharge of millions of gallons of crude oil, e.g. Exxon Valdez ran aground off Alaska in March 1989 discharging oil into the estuary. In February 1996 the Sea Empress ran aground off the coast of Milford Haven in Pembrokeshire, spilling oil into the Cleddau Estuary endangering sea birds and marine life.

- Overhunting and collecting: e.g. for food (bush meat), as exotic pets, fashion, traditional medicine (tiger bone and rhino horn), and for souvenirs and ornaments (turtle shell, ivory).
- Overfishing and agricultural exploitation (see page 52).
- Competition from introduced species: e.g. north American signal crayfish was farmed in the UK but some escaped and now outcompete the native crayfish.

quickfire

(43) List three reasons why species are at risk of extinction.

Conservation

Conservation involves the management of habitats to enhance biodiversity, and is important for a number of reasons:

1. Ethical reasons: we have a responsibility to preserve the environment, not damage it.
2. Possible medical uses: many drugs have been extracted from plants, e.g. quinine from *Cinchona* bark used to treat malaria, some chemotherapy drugs have a plant origin, and presumably many are yet to be discovered.
3. Maintaining a healthy gene pool helps future-proof populations against environmental changes.
4. Agriculture has selectively bred crops from wild varieties. In future, we may need to look at wild varieties to select suitable alleles to grow crops in harsher environments.

Key Term

Conservation: the protection, preservation management and restoration of natural habitats and their ecological communities to enhance biodiversity.

Conservation methods

Legislation:

- Annually, international wildlife trade is estimated to be worth billions of dollars. CITES (the Convention on International Trade in Endangered Species) agreement is enforced by strict customs controls, backed up by fines and even jail sentences, but it is difficult to enforce because not all countries have signed up, and it is very difficult to police or catch the smugglers involved.
- The EU Habitats Directive prevents collection of some birds' eggs, and reduces the picking of wild flowers and overfishing.
- Establish protected areas, e.g. Sites of Special Scientific Interest (SSSIs), and nature reserves, e.g. Gower coast.

Captive breeding programmes in zoos and botanic gardens involve:

- **Seed banks** hold seeds from rare and traditional varieties in controlled environments, to protect against extinction of species.
- **Sperm banks** store sperm from threatened species and are used in captive breeding to ensure genetic variety within populations.
- Rare breed societies maintain older less commercial varieties.
- Species reintroduction has been used successfully following captive breeding programmes to reintroduce species back into the wild, e.g. the red kite has been reintroduced back into mid-Wales.

Key Term

Seed and sperm banks: gene banks, protecting the genes from economically important or threatened plants and animals.

quickfire

(44) List three different conservation methods.

Key Term

Ecotourism: responsible travel to natural areas that conserves the environment and improves the well-being of local people.

Key Term

Monoculture: the cultivation of a single species of crop.

- Education through WWF (World-Wide Fund for Nature) and the Countryside Commission, which are responsible for raising awareness. The Countryside Commission is also responsible for establishing nature reserves.
- **Ecotourism**, e.g. safaris, provides education and raises money to fund local conservation efforts by employing local people. This way, species have more value alive so there are clear incentives to conserve.

Agricultural exploitation

This refers to the way in which food production has had to increase in efficiency and intensity to maximise crop yields in order to feed a growing population. Agricultural exploitation causes conflict between conservation and the need to mass produce food. Following World War II, larger fields were created by removing hedgerows to allow for larger machinery, which led to the loss of habitat for many organisms and reduced biodiversity. Farmers also employed **monoculture** by growing a single species of crop, e.g. wheat, to further increase yields, as all plants required the same nutrients, and harvesting was easier.

Monoculture does have disadvantages:

- It reduces biodiversity as there is only one habitat.
- It provides an ideal environment for pests, so pesticides and herbicides have to be used.
- Farming reduces the flow of recycling of nutrients as when plants die and decompose their constituent elements are returned to the soil, but farmers often remove crop residue and therefore minerals from the soil. Farmers have to add inorganic fertilisers to their fields to increase nutrient content. This can cause eutrophication of waterways.

Farmers are encouraged through the use of subsidies, where they are paid to manage their farms to increase biodiversity.

Deforestation

Causes of deforestation:

- Use of land for agriculture for both subsistence farming and cash crops, e.g.
  - Palm oil
  - Soya bean
  - Biofuels
  - Cattle ranching
- Timber extraction.

Consequences of deforestation:

1. Habitat loss increases causing a reduction in biodiversity.
2. **Soil erosion** increases as tree roots no longer bind soil, so rainfall on exposed slopes can remove top soil.
3. Increase in sedimentation, as top soil is removed from upper slopes and deposited downstream by rivers, increasing the risk of flooding. Soil quality and structure deteriorates as there is no humus added to it from trees.
4. Climate change due to reduced carbon dioxide uptake during photosynthesis.
5. Less transpiration by trees reduces the amount of water vapour returning to the atmosphere, which reduces the amount of rainfall.
6. Loss of plant species and potentially valuable plant chemicals that could be used to treat disease.
7. Desertification.

> **Key Term**
>
> **Soil erosion**: the removal of topsoil, which contains valuable nutrients.

> **quickfire**
>
> ㊺ List four consequences of deforestation.

Managed woodlands

Woodlands can be managed more sustainably by employing:

- Selective cutting, where individual trees are removed, leaving space for remaining tress to grow. The habitat is largely left in place.
- **Coppicing**, which is the process of cutting trees down, allowing the stumps to regenerate for a number of years (usually 7–25) and then harvesting the resulting stems. This encourages great biodiversity in the coppiced woodlands, e.g. wildflowers, grasses and brambles would progressively colonise each new glade as the tree canopy was opened up. The animal species associated with these plants would also then follow.

> **Key Term**
>
> **Coppicing**: cutting down trees close to the ground and leaving them for several years to re-grow.

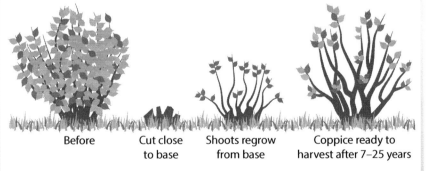

| Before | Cut close to base | Shoots regrow from base | Coppice ready to harvest after 7–25 years |

Coppicing

Native woodland consists mainly of native trees, that is those that have grown there naturally since the last Ice Age and have not been introduced by humans. Native woodland supports a very rich biodiversity, but only 1% of UK woodland is native. The majority of managed woodland in the UK is managed by the Forestry Commission, and is still largely non-native species such as pine.

> **quickfire**
>
> ㊻ Explain the difference between selective cutting and coppicing.

Overfishing

When the harvesting rate is higher than the fish reproductive rate, the fish population size falls. Fishing using nets with a small mesh results in the depletion of younger fish so that the breeding stock is unable to maintain previous population levels.

Commercial fishing using drift netting for pelagic fish, where nets are suspended in the water from floats on the surface, traps non-target species like dolphin and turtles. Trawling for fish in deep water again catches non-target species but is more damaging as it damages the ocean beds, destroying habitats.

Methods to regulate fishing

There are international agreements in place to try to regulate fishing and make it sustainable allowing the population level to be maintained. Methods include:

- Imposing fishing quotas based on scientific estimates of size of fish stocks.
- Enforcing exclusion zones preventing fishing in **overfished** areas.
- Restricting mesh size of nets so only the 'correct' age fish are caught. Larger mesh sizes allow younger fish to escape, survive and reproduce.
- Returning young fish that are caught to sea.
- Forced reduction in fleet size.
- Enforcing fishing seasons (so no fishing in breeding season).
- Allow some fish to return to sea for breeding.
- Encourage fishing of non-traditional varieties.
- Encourage consumer/supermarket to stock ethically fished supplies.

quickfire

㊼ List four legislative strategies used to reduce overfishing.

Fish farming

Fish farming has been seen as a good way to reduce overfishing, and is a better way of producing meat as they convert their food into body protein more efficiently than other animals. Heating of the environment is not required, which lowers the energy input. This has resulted in trout and salmon being routinely farmed in the UK, but there are problems:

- The farms are very densely stocked, meaning that disease spreads more easily, and there is a risk that disease could also spread to wild fish in the vicinity. Antibiotics and pesticides may be overused on the farms, which could lead to antibiotic and pesticide resistance.
- Pesticides used to control fish parasites also harm marine invertebrates.
- The excretory products from the farmed fish are converted into nitrates by bacteria, increasing nitrate concentration in the water, leading to eutrophication.
- Fish can escape and outcompete native species for food.

Sustainability

Environmental monitoring is important in determining whether a development is sustainable. Types of monitoring include:

- Monitoring air quality to identify possible risks to human health.
- Monitoring soil structure, drainage, pH, organic matter and the diversity of living organisms.
- Monitoring water quality in terms of chemical composition, species composition (freshwater invertebrates are a good indicator of pollution, e.g. mayfly nymphs are sensitive to low dissolved oxygen, and cannot survive), and microbial counts.

Environmental Impact Assessments (EIAs) must be performed before new developments are given the go ahead. These were introduced by the EU in 1985 and must consider the environmental consequences of the development. Initially the environment is surveyed so impact upon existing flora and fauna can be considered.

>> *Pointer*

Sustainable development meets the needs of the present without compromising the ability of future generations to meet their own needs.

Planetary boundaries

Nine Earth system processes and their boundaries have been identified which mark the safe zone for the planet. Some of these boundaries have been crossed as a result of human activities: four have been crossed, meaning further events are unpredictable, the crossing of a further two boundaries may have been preventable, and the crossing of one has been avoided.

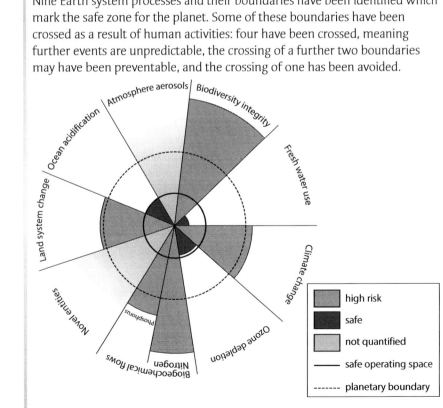

high risk
safe
not quantified
—— safe operating space
------ planetary boundary

The status of planetary systems

Key Term

Planetary boundary: the limits between which global systems must operate to prevent abrupt and irreversible environmental change.

1. Climate change boundary

This is one of two **core boundaries**, and its boundary has been crossed, due to the large volumes of greenhouse gases emitted over the past century. Even drastic reductions in emissions will still only reduce global temperature rise by 2°C, from the 5°C predicted by some scientists. The use of biofuels is seen as a way to achieve this, as by growing them they remove carbon dioxide from the atmosphere during photosynthesis. They are not completely carbon-neutral as energy is used in their production, processing and distribution. Using land for the production of biofuels has it problems: in the UK if all arable land was used to produce biofuels, we wouldn't be able to produce enough to meet our needs and we would have no land to grow food!

Types of **biofuel**:

- Bioethanol is produced from crops such as sugar cane and up to 15% can be added to regular petrol. It is made via simple alcoholic fermentation.

- Biodiesel is widely used in Europe. It is made from vegetable oils, where the fatty acids are reacted with alcohol to produce methyl ester (biodiesel).

- Biogas is methane from the digestion of organic plant and animal wastes. Aerobic digestion of proteins, fats and carbohydrates produces their respective monomers. Methanogenesis from carbohydrate molecules under anaerobic conditions produces methane and carbon dioxide. Methane is naturally produced by decaying organic material at landfill sites, so this can also be collected and used.

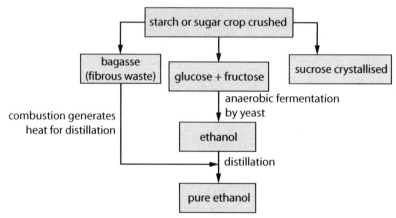

Flow diagram showing bioethanol production

2. The biosphere integrity boundary

Previously known as biodiversity loss and species extinction boundary, it represents the biodiversity of ecosystems, and is the second core boundary to be crossed. Habitat destruction, pollution and climate change are all responsible for reducing biodiversity. It has been estimated that without significant change, more than half of the ocean's species will be lost by the year 2100. The symbiotic relationship between coral and algae is a fragile one: just a 2°C rise in sea temperatures would cause the coral to expel the algae and the coral would die – this is known as coral bleaching.

Key Terms

Core boundary: crossing this planetary boundary would drive the Earth into a new and unpredictable state with severe consequences for the biosphere.

Biofuel: a fuel made by a biological process such as anaerobic digestion, rather than by geological processes that formed fossil fuels.

3. The land-system change boundary

An example of this is deforestation for agriculture, livestock rearing, and the cultivation of biofuel crops. This boundary represents the the misuse of land resulting in too little food being produced.

4. The biogeochemical flows boundary

This boundary refers to the cycling of minerals through cycles like the carbon, phosphorus and nitrogen cycles. Overuse of nitrogen and phosphorus-based fertilisers means that this boundary has already been crossed, and the cycles are no longer self-sustaining.

5. The stratospheric ozone boundary

This boundary represents the destruction of stratospheric ozone by CFCs found in propellants and refrigerants prior to 1987 when the Montreal protocols banned their manufacture following the discovery of a 'hole' in the ozone layer over Antarctica. This action has reversed the crossing of this boundary.

6. The ocean acidification boundary

The pH of the oceans has fallen from 8.16 to 8.03 in the past three centuries which represents a 30% decrease (due to the scale being logarithmic). Fish and invertebrates are particularly sensitive to reductions in pH. Swift action to reduce carbon emissions may prevent this boundary being crossed.

7. The fresh water use boundary

This represents the boundary at which organisms do not have enough regular **fresh water** to survive. Crossing this boundary is avoidable if fresh water use can be reduced. The majority of water on Earth (97%) is saline, and a large proportion of the remaining fresh water is locked up in ice, or is undrinkable due to pollution. Unfortunately, fresh water is not evenly distributed: India has 10% of the world's water but has over 17% of the world's population.

Fresh water availability has decreased due to: irrigation for agriculture, deforestation, water pollution, draining wetlands, increasing population size and per capita (per person) use.

Methods for increasing availability of fresh water:

1. Using water-efficient appliances.
2. Reclaiming waste water for irrigation and industrial use.
3. Stop irrigating non-food crops.
4. Irrigate crops by using drip-irrigation systems.
5. Capture storm water run-off for recharging reservoirs.
6. **Desalinate** salt water.

Desalination works by two main methods: reverse osmosis which separates fresh and seawater by use of a selectively permeable membrane and pressure to force water against its water potential gradient, and the use of solar stills that use the sun's energy to distill water.

> ### Key Terms
>
> **Fresh water**: has a low concentration of dissolved minerals i.e. <0.05% (w/v).
>
> **Desalination**: the removal of salt and other minerals from saline water.

8. The atmospheric aerosol loading boundary

This boundary represents the atmospheric microscopic particles from fossil fuels and dust from quarries. These particles worsen respiratory diseases like asthma, and settle on plant leaves blocking sunlight. This boundary has not been quantified.

9. The introduction of novel entities boundary

Originally called the chemical pollution boundary, it represents pollution from new manufacturing, radioactive materials and nanomaterials. Some chemicals have already been banned due to their toxicity, e.g. PCBs and DDT. The interaction of these chemicals is still relatively unknown, hence it has not been possible to quantify this boundary.

3.7 Homeostasis and the kidney

Key Term

Negative feedback: the mechanism by which the body reverses the direction of change in a system to restore the set point.

Homeostasis is the maintenance of the internal environment within tolerable limits. To accomplish this, the body uses **negative feedback**, whereby the body responds in such a way as to reverse the direction of change. This tends to keep physical parameters constant, e.g temperature at 37°C, and glucose at 90mg per 100cm³ blood. This involves:

1. INPUT – a change away from the set point or norm, e.g. rise in core body temperature.

2. RECEPTOR – a sensor that detects the change from the set point, e.g. temperature receptors.

3. CONTROL CENTRE – or coordinator detects signals from receptors and coordinates a response via effectors, e.g. hypothalamus in the brain.

4. EFFECTOR – bring about changes which returns the body to set point, e.g. glands in skin release sweat.

5. OUTPUT – corrective procedure, e.g. evaporation of sweat cools skin.

>> **Pointer**
An effector is a muscle or a gland.

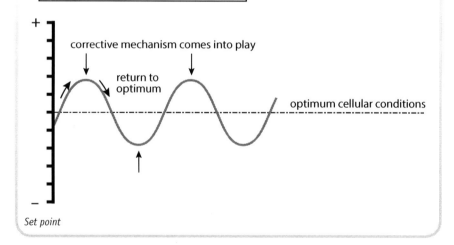

Set point

Excretion

Excretion is the removal of wastes made by the body, e.g. carbon dioxide and water from respiration and urea from the deamination of excess amino acids.

Surplus amino acids are deaminated in the liver: the amine group ($-NH_2$) is removed, converted to ammonia and then into urea. Urea is removed from the body by the kidney. The organic acid that remains can be used in respiration, or converted to lipids or glucose.

quickfire

48 What is the difference between excretion and egestion?

The kidney

The kidney has two main functions:

1. **Excretion** – excretion of nitrogenous waste, i.e. urea from the body.
2. **Osmoregulation** – control of the water potential body fluids including blood.

The body has two kidneys, each containing around a million nephrons, each nephron is 30 mm long. They are supplied with blood containing oxygen and waste (including urea) from the renal artery, and filtered blood returns to the general circulation by the renal vein. Excess water and solutes including urea is called urine, and it drains into the collecting ducts and pelvis which empties urine into the ureter. Each ureter connects to the bladder.

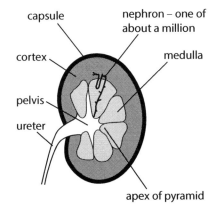

Kidney

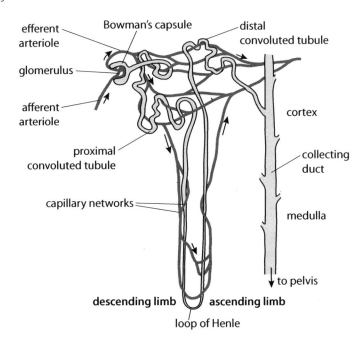

Nephron

Grade boost

It is important to refer to *excess* amino acids, as amino acids are used in protein synthesis – only excess ones are deaminated.

Grade boost

Watch your spelling – ureter and urethra are often confused by candidates due to their similar spelling.

Grade boost

You should be able to identify the region of the kidney from a diagram or micrograph. Remember that the cortex will contain Bowman's capsules but the medulla won't.

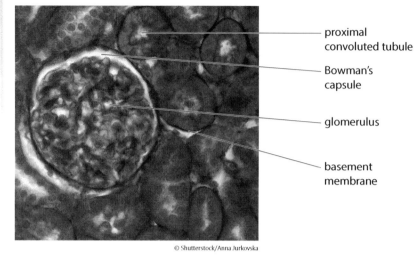

© Shutterstock/Anna Jurkovska

proximal convoluted tubule

Bowman's capsule

glomerulus

basement membrane

Section of kidney cortex seen in light microscope

A network of capillaries surrounds the convoluted tubules and loop of Henle, allowing substances to be reabsorbed into the blood. The capillaries that surround the loop of Henle are referred to as the vasa recta.

Three main processes occur in the nephron:

1. **Ultrafiltration** in the Bowman's capsule where small molecules including water and urea are removed from the blood.

2. **Selective reabsorption** in the proximal convoluted tubule where useful substances such as water, glucose and amino acids are reabsorbed but urea is not.

3. **Osmoregulation** in the loop of Henle and collecting ducts, where the water potential of the blood is regulated.

quickfire

㊾ State two processes carried out by the kidney.

Ultrafiltration

The afferent arteriole is *wider* than the efferent arteriole which creates a *higher* blood pressure than normal in the glomerulus. Substances <68 000 relative molecular mass (rmm) are forced out into Bowman's capsule. This includes glucose, amino acids, salts, water, urea and forms the glomerular filtrate. *Most* proteins in the plasma are >68 000 so remain in the blood with cells: exceptions include HCG hormone which is smaller than 68 000 rmm where its presence can be used to detect pregnancy.

Movement of filtrate is resisted by:

- Capillary epithelium which has pores called fenestrae.

- Basement membrane of Bowman's capsule which acts like a sieve.

- Wall of the Bowman's capsule is made up of highly specialised squamous epithelial cells called podocytes. Filtrate passes between their branches (pedicels).

- Hydrostatic pressure in capsule.

- Low water potential of the blood in glomerulus (lowered by loss of water into capsule).

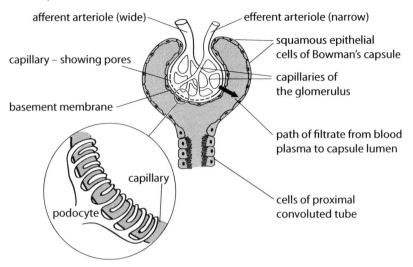

Bowman's capsule

The factors resisting the movement of filtrate determine the filtration rate, which is the rate at which fluid passes from the blood in the glomerular capillaries into the Bowman's capsule. The kidneys receive around 1100 cm³ of blood per minute and produce 125 cm³ of glomerular filtrate in the same time.

Selective reabsorption

Around 85% of filtrate is reabsorbed in the proximal convoluted tubule, which includes all glucose, all amino acids, and most of the water and salts. Urea and *excess* water forms urine. The useful products are reabsorbed In the following ways:

Substance	Method of reabsorption
Mineral ions, and salts	Facilitated diffusion and active transport into epithelial cells
Glucose and amino acids	**Secondary active transport** using a co-transport mechanism with sodium ions. Glucose is co-transported with two sodium ions by facilitated diffusion into the cell. Sodium ions and glucose move separately into the capillaries.
Water	Osmosis
Some filtered proteins and urea	By diffusion

The result is that the filtrate at the end of the proximal convoluted tubule is isotonic with that of the blood plasma.

quickfire

50 List three substances present in the filtrate.

quickfire

51 Explain why HCG is present in the urine but other proteins are not.

quickfire

52 State where you would find Bowman's capsules in the kidney.

quickfire

53 Explain the significance of the diameter of the arterioles supplying the glomerulus.

Key Term

Secondary active transport: the coupling of diffusion, e.g. sodium ions down an electrochemical gradient providing energy for active transport of glucose against its concentration gradient.

Grade boost

When explaining adaptations of cells for selective reabsorption you must *fully* explain your answer, e.g. microvilli provide a large surface area for the absorption of glucose by secondary active transport/co-transport with sodium ions.

The proximal convoluted tubule is adapted for reabsorption in the following ways:

- Cells lining the tubule have a large surface area due to the presence of microvilli and basal channels (infoldings of the membrane in contact with the capillary). There are also large numbers of nephrons.
- Cells contain many mitochondria which provide ATP for active transport of solutes.
- Close association with capillaries which creates a short diffusion pathway between cells and the peritubular capillaries.
- Tight junctions are found between adjacent cells which prevent seepage of reabsorbed materials back into the filtrate.

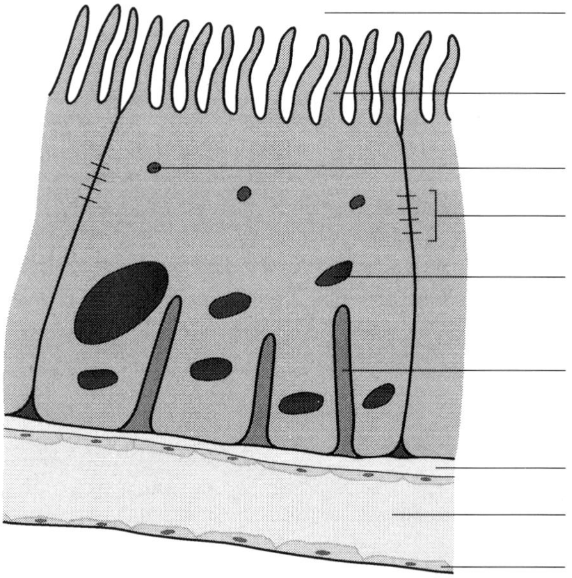

Cuboidal epithelial cell lining proximal convoluted tubule

The glucose threshold refers to the maximum mass of glucose that can be reabsorbed in the proximal convoluted tubule. Where blood glucose concentrations are very high, e.g. in type I and II diabetes, not all of it can be reabsorbed in the tubule, so some remains in the filtrate and therefore in the urine.

The majority of water (90%) is reabsorbed in the proximal convoluted tubule by osmosis. The remainder is reabsorbed in the loop of Henle and distal convoluted tubule and collecting duct. The volume of water reabsorbed in the convoluted tubule and collecting duct varies according to the body's needs.

quickfire

(54) Explain why glucose may be present in the urine of diabetic patients.

Osmoregulation

Controlling the water potential of the body is important in animals as it maintains the concentrations of enzymes and metabolites, and prevents cells from bursting or crenating.

The loop of Henle is responsible for reabsorption of some water from the descending limb, but its main function is to create an increasing ion concentration within the interstitial region of the medulla, which allows water to be reabsorbed by the collecting duct. The volume of water reabsorbed from the distal convoluted tubule and collecting duct, and hence the resulting water potential of the blood, is influenced by antidiuretic hormone, which increases the permeability of the tubule and duct walls to water.

 Grade boost

Longer loops of Henle are found in mammals that have evolved in dry habitats. More concentrated urine can be produced because more Na^+ and Cl^- ions can be actively transported out of the descending limb.

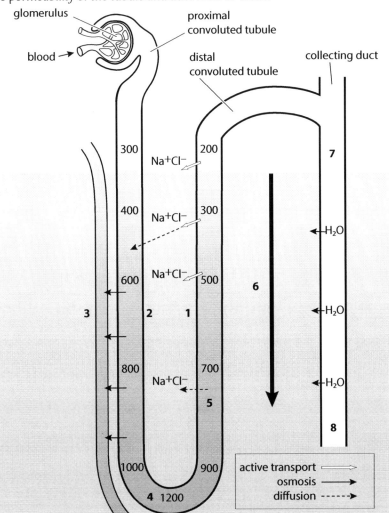

Concentrations shown in mOsm kg^{-1}

Counter current multiplier

1. Na^+ and Cl^- ions are actively pumped out of the ascending limb.

2. This creates an increasing ion concentration in the interstitial region.

3. Walls of the descending limb are permeable to water so water leaves by osmosis into the interstitial region before entering the capillaries (vasa recta).

4. Water is progressively lost down the descending limb reaching typically a concentration of 1200 mOsm kg^{-1} of water at the base. (Longer loops can reach much higher concentrations because more Na^+ and Cl^- ions can be actively transported out of the ascending limb.)

5. The concentration of the filtrate decreases in the lumen of the nephron in the ascending limb, as Na^+ and Cl^- ions are actively pumped out.

6. This creates an increasing ion concentration gradient in the interstitial region towards the base of the loop.

7. Water passes out of collecting duct by osmosis into the interstitial region before entering the vasa recta.

8. As water passes out of filtrate in the collecting duct, the concentration of the filtrate increases, but it is always lower than the fluid in the interstitial region of the medulla, so water will continue to leave by osmosis. The two liquids flow in opposite directions past each other, resulting in a greater exchange of substances between them than if they flowed in the same direction. This is referred to as a counter current multiplier. This ensures that the concentration of the filtrate is always lower than the interstitial fluid in the medulla.

Role of ADH in osmoregulation

The water potential of the blood varies when animals become:

1. Over-hydrated due to excess water intake, or low salt intake.

2. Dehydrated due to low water intake, sweating or high salt intake.

The water potential of the blood is controlled by receptors called osmoreceptors in the hypothalamus, which respond by triggering the release of more or less **antidiuretic hormone** (ADH) into the blood from the posterior lobe of the pituitary gland. Osmoregulation is controlled by negative feedback.

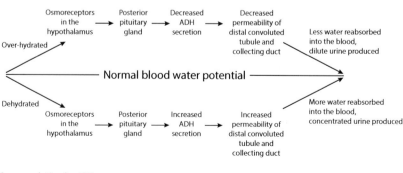

Osmoregulation by ADH

ADH binds to membrane receptor proteins found on the surface of cells lining the distal convoluted tubule and collecting duct walls. The binding of ADH triggers vesicles containing intrinsic membrane proteins called aquaporins containing pores that allow water to move, to fuse with the cell membrane. The aquaporins allow water to pass through the walls down the water potential gradient. When ADH concentration falls, aquaporins are removed from the cell membrane.

3.7

The table shows the typical concentrations of three solutes in the nephron.

Solute	Mean concentration / g dm⁻³		
	Bowman's capsule	Proximal convoluted tubule	Distal convoluted tubule
Glucose	0.12	0.00	0.00
Urea	0.35	0.65	6.25
Sodium ions	0.28	0.24	0.02

a) Explain the change in urea concentration between the Bowman's capsule and the distal convoluted tubule.

b) What can be concluded about the reabsorption of sodium ions?

c) Suggest why diabetic patients can suffer damage to the cells lining the distal convoluted tubule.

Kidney failure

The main causes of kidney failure are diabetes, high blood pressure, auto-immune disease, infection, and crushing injuries. Kidney disease is treated by balancing body fluids using:

- Medication to control blood potassium and calcium levels, which can lead to heart disease and kidney stones if not controlled.
- A low protein diet to reduce concentration of excess amino acids, and hence concentration of urea.
- Drugs to lower blood pressure, e.g.
  - Beta blockers which, reduce the effect of adrenalin.
  - Calcium channel blockers, which dilate blood vessels lowering blood pressure.
  - ACE inhibitors, which reduce the effect of angiotensin. Angiotensin causes blood vessels to constrict.
- Dialysis involves using a dialysis fluid that contains glucose at the same concentration as the blood, but has no urea and a low ion concentration. The result is that urea, some ions and water diffuse out of the blood, but glucose remains. There are two types (see below).
- Kidney transplant, as a final resort for end-stage renal disease. It involves transplanting one kidney from a donor who is closely matched to ensure tissue compatibility. Immunosuppressants have to be used afterwards to prevent organ rejection.

Dialysis

There are two types in use:

1. Haemodialysis which takes blood (usually from an artery in the arm), and passes it through a dialyser containing thousands of fibres each consisting of selectively permeable dialysis tubing and dialysis fluid. To ensure maximum transfer, a counter current is used where blood and dialysis fluid move in opposite directions. Heparin is used to prevent the blood from clotting. Dialysis takes several hours and is repeated several times each week.

2. Peritoneal dialysis involves passing dialysis fluid into the peritoneum through a catheter. The peritoneum contains numerous capillaries which exchange materials with the dialysis fluid, which is changed after about 40 minutes, and the process repeated several times a day. This type has the advantage that the patient is able to move around, but it is less effective than haemodialysis so fluid retention is likely.

Excretion in other animals

Freshwater fish excrete ammonia:

- Ammonia is highly soluble but is very toxic so it cannot be stored. It must be excreted immediately using large volumes of water to dilute it (which is freely available to freshwater fish).

Mammals excrete urea:

- Urea is much less toxic than ammonia and so requires less water to dilute it, and can be stored for short periods of time.
- It requires energy to convert ammonia to urea but is an adaptation to life on land, as it helps prevent dehydration because less water is needed to excrete it.

Birds, reptiles and insects excrete uric acid:

- Uric acid is virtually non-toxic and therefore requires very little water to dilute it.
- The conversion of ammonia to uric acid requires much energy but allows these animals to survive in very arid environments.
- A major advantage for birds is that very small volumes of water are needed reducing weight in flight.

The length of the loop of Henle is an adaptation to where the animal has evolved. Beavers, which live in freshwater where water is plentiful, have very short loops of Henle and produce large volumes of dilute urine. Animals that live in arid environments, e.g. the kangaroo rat, have much longer loops of Henle and produce small volumes of highly concentrated urine. They have a higher proportion of these nephrons which are referred to as juxtamedullary nephrons, with the Bowman's capsule being located closer to the medulla, and loops of Henle, which penetrate deep into the medulla. The longer the loop of Henle, the more concentrated the urine that can be produced (which saves water), because a higher ion concentration in the medulla can be created by the counter current multiplier.

Animals in arid environments also rely more on metabolic water, e.g. the camel, which largely respires fat stored in its hump. Other mammals show behavioural adaptations such as being nocturnal, coming out at night to forage when it is cooler.

quickfire

57 Explain why freshwater fish excrete ammonia but birds excrete uric acid.

quickfire

58 Explain two adaptations that mammals like camels show for life in arid environments.

3.8 The nervous system

Hormonal control changes things more slowly, involves a long-term response, and relies upon chemicals being carried by the blood. Nervous control changes things more rapidly, involves a short-term response and relies upon information carried by neurones. The nervous system is responsible for detecting changes within the internal or external environment (a stimulus), processing that information and initiating a response. This is achieved through the stimulus response model:

Stimulus → Detector → Coordinator → Effector → Response

- Stimulus is a change in the environment.
- A detector contains cells which can detect stimuli, e.g. visible light by the retina, sound by the inner ear, pressure by Pacinian corpuscles in the dermis of the skin, temperature by the dermis of the skin, or chemicals through taste and smell. It converts energy from one form, e.g. light, into an electrical impulse.
- The coordinator is the central nervous system consisting of the brain and spinal cord. It coordinates the response.
- An effector brings about a response. It is either a muscle or a gland.
- The response is the change in the organism.

In humans, the nervous system consists of the central nervous system (CNS) and the peripheral nervous system. The peripheral nervous system is made up of:

1. Somatic nervous system consisting of pairs of nerves that originate from the brain and spinal cord, containing both sensory and motor neurones.
2. Autonomic nervous system which controls involuntary actions, e.g. digestion and control of heartbeat.

Neurones

There are three types of neurones in humans:

1. Sensory neurones that carry impulses from receptors to the CNS.
2. Relay or connector neurones within the CNS that receive impulses from sensory or other relay neurones and transmit them onto motor neurones.
3. Motor neurones that transmit impulses from the CNS to effectors (muscles or glands).

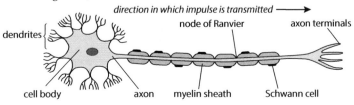

direction in which impulse is transmitted ⟶

dendrites — node of Ranvier — axon terminals — cell body — axon — myelin sheath — Schwann cell

Mammalian motor neurone

quickfire

(59) Which neurones carry impulses towards an effector?

Grade boost

You should be able to draw and label a motor neurone and describe the function of its components.

Structure	Function
Cell body (centron)	Contains a granular cytoplasm with ribosomes for protein synthesis. DNA is present within a nucleus and acts as the site for transcription.
Axon	Carries the impulse away from the cell body.
Myelin sheath	Surrounds the axon (and dendron in sensory neurones) providing electrical insulation resulting in faster impulse transmission.
Schwann cell	Surrounds the axon (and dendron in sensory neurones) and forms the myelin sheath.
Nodes of Ranvier	Gaps in the myelin sheath between Schwann cells are approximately 1μm wide where the axon membrane is exposed. They allow faster nerve impulse conduction (saltatory conduction).
Axon endings	Secrete a neurotransmitter which results in depolarisation of the adjacent neurone.
Synaptic end bulbs	Swelling found at the end of an axon where the neurotransmitter is synthesised.

Table showing the functions of the main components of a motor neurone

Reflex arc

Reflexes are rapid, automatic responses to stimuli that could prove harmful to the body, and are therefore protective in nature. In a reflex arc a stimulus is detected by the receptor and passed to the CNS along a sensory neurone. The impulse is then relayed directly to a motor neurone and its effector by a relay neurone. The response is rapid and involves the contraction of a muscle or release of a hormone. In most cases a reflex involves the spinal cord, but some reflexes, e.g. pupil reflex, will involve the brain as it is the closest part of the CNS.

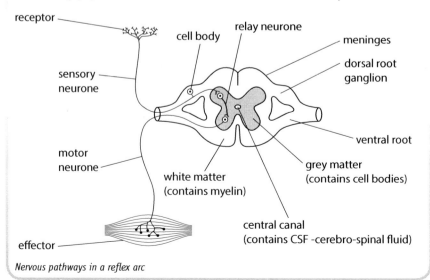

Nervous pathways in a reflex arc

quickfire

60 Why is the white matter white in colour?

quickfire

61 Which neurones in the spinal cord are unmyelinated?

Nerve nets

Simple animals, e.g. Cnidarians like *Hydra*, do not possess a nervous system like mammals, but have a simplified nervous system called a nerve net. It consists of sensory photoreceptors and touch receptors in the wall of the body and tentacles. Ganglion cells provide connections between the neurones in several directions but they do not form a brain.

Cnidarian nerve net	Mammalian nervous system
1 type of simple neurone	3 types of neurone (sensory, relay and motor)
Unmyelinated	Myelinated
Short, branched neurones	Long, unbranched neurones
Impulse transmitted in both directions	Impulse transmitted in one direction
Slow impulse transmission	Fast impulse transmission

Grade boost

When distinguishing between Cnidarians and mammals, a table helps to show comparative statements.

Nerve net in Hydra

quickfire

⑥ How are neurones in Cnidarians different from mammals?

The nerve impulse

quickfire

⑥⑤ What is the potential difference across the membrane of a neurone at rest?

When a neurone is at rest, i.e. no impulses are being transmitted, it is said to be at **resting potential**. At rest, the charge across the axon membrane is slightly negative at around −70 mV with respect to the inside (it is more positive outside). Resting potential is created because:

- Phospholipid bilayer is impermeable to Na^+/K^+ ions.
- These ions are only able to move across the membrane through intrinsic proteins and the sodium/potassium pump (active transport).
- Some intrinsic proteins have 'gates', which can be opened or closed to allow/inhibit ion movement.
- Na^+ gates allow Na^+ ions to pass in, K^+ gates allow K^+ ions to pass out.
- Most K^+ gates are OPEN whereas most Na^+ gates are closed. This makes the membrane 100 times more permeable to K^+ ions than Na^+ ions.
- The resting potential is always negative, because there are fewer positive ions inside than outside.

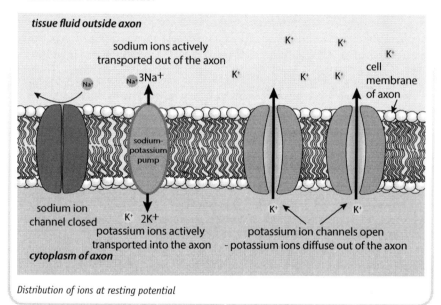

Distribution of ions at resting potential

The action potential

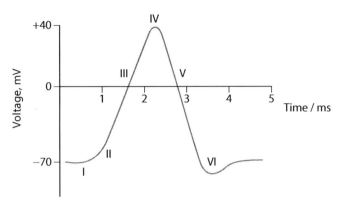

Action potential

I   At resting potential, the Na⁺ gates are closed and some K⁺ gates are open which together with the Na⁺ /K⁺ pump result in a potential difference (charge) across the membrane of −70 mV.

II Energy of stimulus arriving causing Na⁺ voltage-gated gates to OPEN and Na⁺ ions flood in down their concentration gradient, **depolarising** the neurone. Now the charge across the membrane becomes MORE positive due to MORE positive charges inside.

III As more Na⁺ ions enter, more gates open so even more Na⁺ ions rush in *(positive feedback)*.

IV When potential reaches +40 mV the neurone is depolarised. Na⁺ gates close preventing further influx of Na⁺ ions. K⁺ gates then begin to open.

V   K⁺ ions flood out of the neurone down their concentration gradient lowering the positive (+) gradient across the membrane. As a result, further K⁺ channels open, resulting in even more K⁺ ions leaving the neurone. The neurone becomes repolarised.

VI Too many K⁺ ions leave the neurone so the electrical gradient overshoots −70 mV reaching around −80 mV (which is called hyperpolarisation). To re-establish the resting potential (−70 mV), K⁺ gates now close, and the Na⁺ /K⁺ pump re-establishes the resting potential.

quickfire

⑭ Match the numbers on the action potential graph to the following states:
A Depolarised
B Resting potential
C Hyperpolarised
D Depolarising
E Repolarising

Key Terms

An action potential: the rapid rise and fall of the electrical potential across a neurone membrane as a nerve impulse passes.

Depolarisation: a temporary reversal of the potential difference across the membrane of a neurone such that the inside becomes less negative relative to the outside, as an action potential is transmitted.

Impulse propagation

The following occurs in unmyelinated neurons:

- ☐ Polarised
- ☐ Depolarised
- ☐ Repolarised

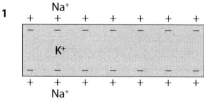

1. Neurone membrane is POLARISED, i.e. charge across membrane is −70 mV.

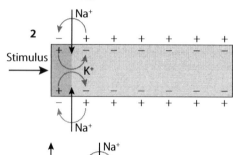

2. Na⁺ channels OPEN so Na⁺ ions rush into axon cytoplasm. A local circuit is established where Na⁺ ions are pumped out of the adjacent Na⁺/K⁺ pump.

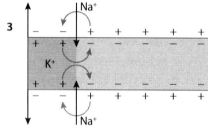

3. Na⁺ channels in the adjacent part of membrane OPEN causing depolarisation.

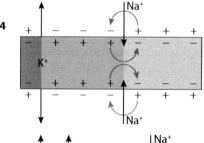

4. Meanwhile, Na⁺ channels CLOSE, K⁺ channels OPEN which causes repolarisation behind it.

5. Depolarisation continues along neurone membrane. The membrane at the start is now POLARISED again.

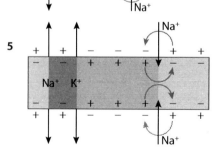

In myelinated neurons, ions can only move across the membrane at the nodes of Ranvier where there is no myelin present, therefore local circuits are established over greater distances (between each node). Depolarisation only occurs at the nodes and the action potential effectively 'jumps' from node to node, increasing the speed of impulse transmission.

Action potential

Grade boost

Remember it is the action potential that 'jumps' from node to node, NOT the impulse.

Refractory period

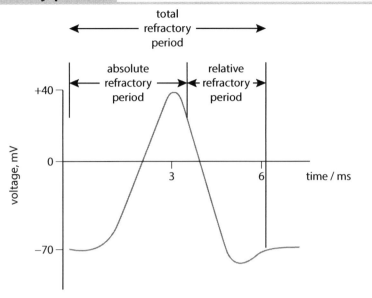

Refractory period

- *Total refractory period* lasts for approximately 6 ms, and represents the period during which it is not normally possible to send another impulse.
- The *absolute refractory period* is the period during which it is NOT possible to send another impulse, irrespective of how BIG the stimulus is.
- The *relative refractory period* is the period during which it is possible to send another impulse, if the stimulus is big enough to overcome the THRESHOLD.

All or nothing rule

Impulses will pass if a threshold value is exceeded (usually −55 mV). A large stimulus will result in more impulses passing per second (increased frequency of action potentials) rather than a greater level of depolarisation. Impulses either pass or do not, and they are always the same size, which is referred to as the all or nothing law.

quickfire

65 What is the significance of the refractory period?

≫ *Pointer*

Some motor neurones in mammals are unmyelinated, e.g. motor neurone to a gland.

quicKfire

66 State two factors that affect the rate of impulse transmission in mammals.

Factors affecting the speed of impulse transmission

1. Myelination: saltatory conduction is faster than impulse transmission in unmyelinated neurones, as depolarisation only occurs at the nodes of Ranvier (that occur every 1mm or so along the axon length) so the action potential effectively 'jumps' from node to node. The rate of transmission varies from 1 m/s, in unmyelinated neurones to 100 m/s in myelinated ones.

2. Diameter of axon: impulse transmission speed increases with axon diameter due to less leakage of ions from larger axons (due to a larger volume to surface area).

3. Temperature – impulse transmission speed increases with temperature because the rate of diffusion increases due to the increased kinetic energy of ions involved, but only in organisms which do not control their internal body temperature (some ectotherms).

The synapse

A chemical synapse exists as a 20 nm gap between two neurones. The impulse is transmitted from one to the other by a neurotransmitter, which diffuses across the synaptic cleft from the pre-synaptic membrane to receptors on the post-synaptic neurone, triggering depolarisation in the post-synaptic neurone. An example of a neurotransmitter used by the parasympathetic nervous system is acetylcholine.

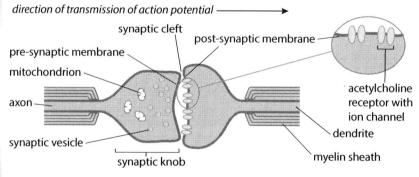

A synapse

Synapses have a number of functions. They:

- Transmit information between neurones.
- Transmit information in one direction only.
- Act as junctions.
- Filter out low-level stimuli.
- Prevent over stimulation of neurone and fatigue.

Synaptic transmission

The events can be summarised as follows:

- An impulse arrives at the pre-synaptic knob.
- Calcium channels open, causing calcium ions to diffuse rapidly into the pre-synaptic knob.
- The vesicles containing the neurotransmitter acetylcholine migrate to and fuse with the pre-synaptic membrane.
- Contents of the vesicles are released into the synaptic cleft by exocytosis.
- Acetylcholine molecules diffuse across the cleft and bind to receptors on the post-synaptic membrane causing sodium channels to open.
- Na^+ ions rush into the post-synaptic neurone resulting in depolarisation of the post-synaptic membrane. An action potential is initiated.
- (Acetyl) Cholinesterase splits the acetylcholine into ethanoic acid and choline, releasing them from the receptor, and sodium channels close. The products diffuse back across the cleft.
- Products are reabsorbed into the pre-synaptic knob.
- ATP is used to reform acetylcholine in the pre-synaptic knob.

Repeated depolarisation of the post-synaptic neurone is prevented by:
- The hydrolysis of acetylcholine.
- Reabsorption of ethanoic acid and choline back into the pre-synaptic knob.
- Active transport of calcium ions out of the pre-synaptic knob, which prevents further exocytosis of neurotransmitter.

If insufficient acetylcholine is released, not enough sodium channels open on the post-synaptic membrane to exceed the threshold potential of −55 mV, so an action potential is not initiated.

quickfire

67 Name the type of transport involved when calcium ions move into the pre-synaptic knob.

quickfire

68 Explain the large number of mitochondria present in the pre-synaptic knob.

Effects of chemicals on synapses

Drugs have two main types of effects:

1. Excitatory (stimulants or agonists) like caffeine and cocaine which result in more action potentials.

2. Sedatives (inhibitory) like cannabis which result in fewer action potentials.

Organophosphorus insecticides act as agonists by inhibiting cholinesterase, so acetylcholine lingers at the synapse causing repeated depolarisation of the post-synaptic membrane.

Many drugs e.g. nicotine, mimic the action of neurotransmitters, but unlike acetylcholine, nicotine is not removed by the action of cholinesterase. Over time the body produces less acetylcholine, and the person becomes more reliant on nicotine for the normal functioning of the synapse. Nicotine also causes the increased release of dopamine in the brain, resulting in addiction.

Unit 3 Summary

3.1 Importance of ATP

- ATP belongs to the group of molecules called nucleotides.
- There are three types of phosphorylation: oxidative, photophosphorylation and substrate level phosphorylation.
- ATP yields 30.6 kJ of energy when the high energy bond is broken. This energy can be used for many processes including DNA and protein synthesis.
- ATP is the universal energy currency: all reactions in all organisms.

3.2 Photosynthesis

- The leaf is adapted for photosynthesis.
- Chloroplast pigments act as transducers.
- Photosynthetic pigments include chlorophyll a and b, xanthophylls and β carotene each absorbing different wavelengths of light.
- Light is harvested by antenna complexes.
- Photophosphorylation occurs via two pathways: cyclic and non-cyclic.
- The light-dependent stage occurs in the thylakoid membrane of the chloroplast.
- Non-cyclic photophosphorylation yields reduced NADP and 2 molecules of ATP as electrons are passed through a series of carriers.
- Water undergoes photolysis releasing oxygen and 2 electrons to photosystem II.
- The Calvin cycle takes place in the stroma, and fixes one carbon dioxide molecule per turn of the cycle via the enzyme RuBisCO. Glycerate-3-phosphate is reduced to triose phosphate by reduced NADP.
- A limiting factor is one that affects the rate of a reaction.

3.3 Respiration

- Aerobic respiration occurs in the mitochondria and yields a theoretical maximum 38 ATP.
- Glycolysis occurs in the cytoplasm producing 2 ATP (net) and 2 reduced NAD.
- A series of dehydrogenation and decarboxylation reactions occur during the link reaction and Krebs cycle within the mitochondrial matrix producing 2 ATP molecules and reduced NAD and reduced FAD, and carbon dioxide.
- Most ATP is produced in the electron transport chain by oxidative phosphorylation using electron energy to pump protons across the inner membrane into the inter-membrane space, thus creating a proton gradient.
- ATP is made by ATP synthetase as protons flow back into the matrix by chemiosmosis.
- Anaerobic respiration yields less ATP than aerobic respiration, and produces lactate in animals, ethanol and carbon dioxide in plants and yeast.

3.4 Microbiology

- Bacteria are classified according to their shape, cell wall structure, and their metabolic, antigenic and genetic features.
- Gram-positive bacteria have a thick layer of peptidoglycan and so retain the crystal violet stain appearing purple.
- Gram-negative bacteria do not retain the crystal violet stain.
- Viable cell counts estimate the number of living cells, whereas total cell counts estimate the total number of cells both living and dead.
- Samples for a viable cell count require a serial dilution to produce results that are countable.

3.6 Human impact on the environment

- Species are at risk of extinction through changes in climate and habitat loss.
- It is important to conserve species.
- Legislation and captive breeding programmes are used to conserve species.
- Agricultural exploitation produces a conflict between mass production of food and conservation.
- Deforestation creates land for agriculture and timber, and causes soil erosion leading to desertification.
- Woodlands can be managed by selective cutting and coppicing.
- Overfishing reduces biodiversity and damages habitats.
- There are nine planetary boundaries, several of which have been or are close to being breached.

3.5 Population size and ecosystems

- Abundance is a measure of how many individuals exist within a habitat.
- Mark-release-recapture can be used to estimate total population size.
- Quadrats and transects can be used to estimate percentage cover of plant species.
- Food chains are limited in length due to loss of energy at each stage.
- Energy flow through ecosystems can be calculated.
- Primary succession is from previously uncolonised areas, e.g. bare rock, whilst in a secondary succession on land, soil already exists.
- Carbon is recycled by decomposers (microorganisms) in the carbon cycle. Nitrogen is recycled by the nitrogen cycle via ammonification, nitrification, nitrogen fixation and denitrification by different species of bacteria.
- The excess use of fertilisers leads to eutrophication and algal blooms.

3.7 Homeostasis and the kidney

- Homeostasis is the maintenance of the internal environment within tolerable limits.
- Negative feedback is the mechanism by which the body reverses the direction of change in a system to restore the set point.
- Excretion is the removal of wastes, e.g. carbon dioxide and water, made by the body.
- Ultrafiltration in the Bowman's capsule where small molecules including water and urea are removed from the blood.
- Selective reabsorption in the proximal convoluted tubule where useful substances such as water, glucose and amino acids are reabsorbed but urea is not.
- Osmoregulation in the loop of Henle and collecting ducts, where the water potential of the blood is regulated.
- The water potential of the blood is controlled by osmoreceptors in the hypothalamus, which respond by triggering the release of more or less antidiuretic hormone (ADH) into the blood from the posterior lobe of the pituitary gland.
- Kidney failure can be treated by medication, dialysis or transplant.
- Different animals excrete different nitrogenous waste: freshwater fish excrete ammonia, mammals excrete urea, and birds, reptiles and insects excrete uric acid.
- The length of the loop of Henle is correlated with where an animal evolved.

3.8 The nervous system

- In humans, the nervous system comprises the central nervous system (CNS) and the peripheral nervous system.

- Sensory neurones that carry impulses from receptors to the CNS.

- Relay or connector neurones within the CNS that receive impulses from sensory or other relay neurones and transmit them to motor neurones.

- Motor neurones transmit impulses from the CNS to effectors (muscles or glands).

- Reflexes are rapid, automatic responses to stimuli that could prove harmful to the body, and are therefore protective in nature.

- Simple animals, e.g. Cnidarians like Hydra do not possess a nervous system like mammals, but have a simple nervous system called a nerve net.

- When a neurone is at rest, i.e. no impulses are being transmitted, it is said to be at resting potential. At rest, the charge across the axon membrane is negative at around −70 mV with respect to the inside.

- An action potential is the rapid rise and fall of the electrical potential across a neurone membrane as a nerve impulse passes.

- The total refractory period lasts for approximately 6 ms, and represents the period during which it is not normally possible to send another impulse.

- The absolute refractory period is the period during which it is not possible to send another impulse, irrespective of how big the stimulus is.

- The relative refractory period is the period during which it is possible to send another impulse, if the stimulus is big enough to overcome the threshold.

- Impulses pass if the threshold value is exceeded (usually −55 mV).

- The speed of impulse transmission is affected by myelination, the diameter of the axon and temperature.

- Synapses transmit impulses between neurones, are unidirectional, act as junctions and filter out low level stimuli.

- Drugs act as either stimulants or sedatives by influencing neurotransmitters at synapses.

Unit 4 Knowledge and Understanding

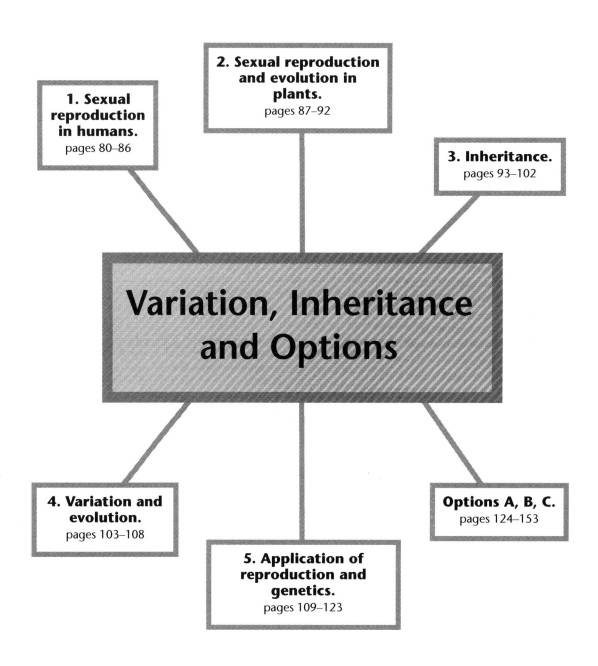

1. Sexual reproduction in humans.
pages 80–86

2. Sexual reproduction and evolution in plants.
pages 87–92

3. Inheritance.
pages 93–102

Variation, Inheritance and Options

4. Variation and evolution.
pages 103–108

5. Application of reproduction and genetics.
pages 109–123

Options A, B, C.
pages 124–153

4.1 Sexual reproduction in humans

Male reproductive system

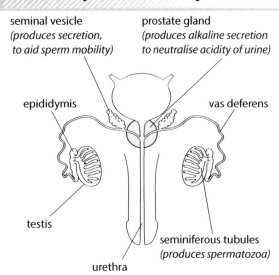

seminal vesicle
*(produces secretion,
to aid sperm mobility)*

prostate gland
*(produces alkaline secretion
to neutralise acidity of urine)*

epididymis

vas deferens

testis

seminiferous tubules
(produces spermatozoa)

urethra

Male reproductive system

- Each testis contains around one thousand seminiferous tubules where spermatozoa are formed.
- Spermatozoa collect in the epididymis where their motility improves.
- Seminal vesicles secrete mucus and prostate fluid mixes with spermatozoa during ejaculation.
- These fluids maintain sperm mobility, provide nutrients, e.g. fructose, and are alkaline which neutralises acidity found in urine and the vagina.
- The resultant fluid containing sperm is called semen, and leaves the penis by the urethra.

Female reproductive system

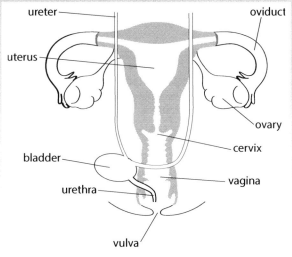

ureter

uterus

oviduct

ovary

cervix

bladder

vagina

urethra

vulva

Female reproductive system

- There are two ovaries where oocytes mature from germinal epithelial cells.
- Each month, one secondary oocyte is released during ovulation from the surface of one of the ovaries.
- Cilia lining the fallopian tube (oviduct) waft the secondary oocyte along.
- The uterus consists of a thin outer layer called the perimetrium. Inside this is the muscle layer or myometrium.
- The endometrium is the innermost layer consisting of a mucous membrane which is well supplied with blood. This layer is shed each month during the menstrual cycle if an embryo fails to implant.

 **Grade boost**

Watch your spelling. Ureter and urethra are easily confused, and it is prostate gland NOT prostrate!

Gametogenesis

Gametogenesis is the production of gametes through a series of mitotic and meiotic divisions in the testis and ovaries:

1. Sperm are produced by spermatogenesis.
2. Eggs are produced by oogenesis.

Spermatogenesis

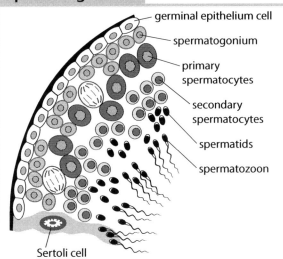

- germinal epithelium cell
- spermatogonium
- primary spermatocytes
- secondary spermatocytes
- spermatids
- spermatozoon
- Sertoli cell

Illustration of a seminiferous tubule

- As you move from the outside towards the centre of the seminiferous tubule, the cells become more mature.
- Diploid germinal epithelial cells divide by means of mitosis to produce diploid spermatogonia.
- Primary spermatocytes (2n) divide by meiosis I to produce secondary spermatocytes (n).
- Secondary spermatocytes (n) undergo meiosis II to make spermatids (n).
- Spermatids differentiate and mature into spermatozoa (n).
- Sertoli cells provide spermatozoa with nutrients and protect them from the male's immune system.
- Interstitial cells secrete testosterone.

 Grade boost

You could be asked to compare spermatogenesis and oogenesis or spermatozoon and secondary oocyte structure.

》 Pointer

Remember diploid = 2n, haploid = n, where n is number of chromosomes.

Grade boost

Ovulation is the release of a secondary oocyte – not an ovum.

Oogenesis

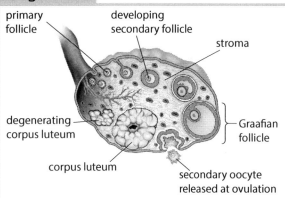

- primary follicle
- developing secondary follicle
- stroma
- degenerating corpus luteum
- Graafian follicle
- corpus luteum
- secondary oocyte released at ovulation

Illustration of an ovary

- Before birth, germinal epithelial cells divide by means of mitosis to produce oogonia (2n) and then primary oocytes (2n).
- Primary oocytes are surrounded by germinal epithelial cells which form the primary follicle.
- Primary oocytes begin meiosis I but stop at prophase I. Division resumes from puberty.
- Each month, a primary oocyte continues meiosis I to produce a secondary oocyte and a polar body, both of which are haploid.
- The primary follicle also develops into a secondary follicle, which matures into a Graafian follicle. This migrates to the surface and bursts releasing the secondary oocyte (ovulation).
- The secondary oocyte undergoes meiosis II, stopping at metaphase II.
- If a sperm meets the secondary oocyte and enters, meiosis II is completed resulting in the production of the ovum (n) and second polar body (n).
- The sperm pronucleus can now fuse with the ovum pronucleus to produce a diploid zygote.
- Following fertilisation, the Graafian follicle becomes the corpus luteum and produces progesterone. If no fertilisation occurs it regresses.

 **Link** Revisit meiosis in AS.

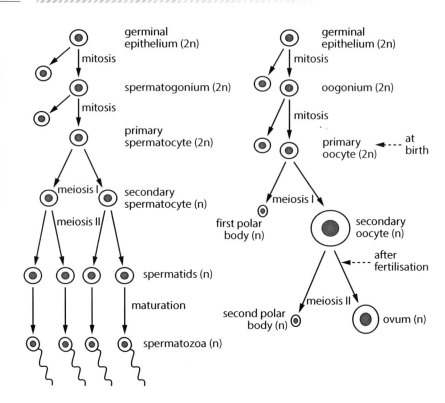

Flow diagrams of gametogenesis

quickfire

① Name the structures that produce spermatozoa.

quickfire

② Name two structures that produce secretions to aid sperm mobility.

Structure of a human spermatozoon

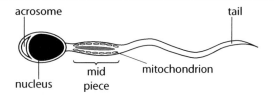

Diagram of a spermatozoon

- Each spermatozoon head is approximately 5 µm long, with a 50 µm tail.
- Within the head is a haploid nucleus, and the acrosome which contains proteases to digest the cells of the corona radiata and zona pellucida.
- Mid piece contains many mitochondria which provide ATP for movement.
- The tail (flagellum) moves in a circular wave motion to propel the spermatozoon forward.

quickfire

③ What is the function of the mitochondria in the mid piece of a spermatozoon?

Structure of a human secondary oocyte

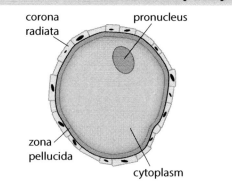

Diagram of a secondary oocyte

- A typical ovum measures 120 µm in diameter, and is one of the largest cells in the human body.
- Fats and albumins contained within the cytoplasm provide nutrition for the developing embryo until it implants in the wall of the uterus and the placenta can provide nutrients.
- Changes to the zona pellucida following entry of a single spermatozoon prevent polyspermy (entry of additional sperm).

The human menstrual cycle

The human menstrual cycle takes about one month to complete. The process is controlled by gonadotrophic hormones from the anterior pituitary and hormones from the ovary itself. From about day 5 of the cycle, FSH (follicle-stimulating hormone) is released from the anterior pituitary promoting maturation of the Graafian follicle stimulating production of the steroid hormone oestrogen by the ovary. Oestrogen has the effect of increasing the thickness and vascularity of the uterus lining, the endometrium, in preparation for the implantation of a fertilised ovum.

By about day 14 oestrogen levels are high enough to inhibit further production of FSH by negative feedback and stimulate release of luteinising hormone (LH). The sudden release of LH induces ovulation. It also promotes the formation of a corpus luteum and stimulates the release from it of another steroid hormone, progesterone. High levels of progesterone building up over the next 10 days inhibit both FSH and LH. Oestrogen and progesterone levels drop and the endometrial lining breaks down resulting in menstruation. If fertilisation has occurred, progesterone levels will remain high and this will inhibit the release of FSH and LH by the pituitary.

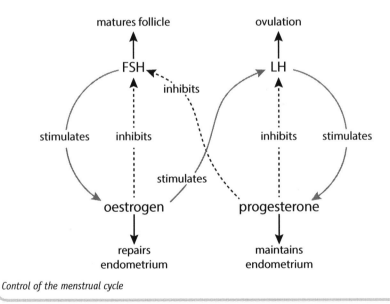

Control of the menstrual cycle

 Grade boost

Learn the names of the different cells in spermatogenesis and oogenesis.

quickfire

④ What is the function of:
a) The Sertoli cells?
b) The interstitial cells?

extra

2.3

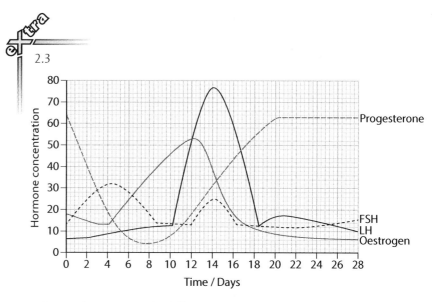

Use the graph to answer the following questions:

a) What evidence is there that FSH stimulates the release of oestrogen?

b) Explain when ovulation is most likely to occur.

c) Explain which ovarian hormone could be injected to stimulate ovulation.

d) Explain which ovarian hormone could be injected to prevent ovulation.

Sexual intercourse

During sexual intercourse, the erect penis is inserted into the vagina, and following movements of the penis, semen is ejaculated into the vagina by contractions of the smooth muscle in the walls of the epididymis, vas deferens and penis. The force of ejaculation propels some sperm through the cervix and into the uterus, with the remainder collecting at the top of the vagina.

Events that lead to fertilisation

Sperm respond to chemicals produced by the oocyte and begin to swim through the uterus and into the oviduct. Once there, sperm can remain viable for a few days, but are most fertile in the 12–24 hours following intercourse. The oocyte remains viable for only 24 hours following ovulation, so needs to be fertilised relatively quickly after ovulation. The events that lead to fertilisation are:

1. Cholesterol and glycoproteins are removed from the cell membrane covering the sperm's acrosome making the membrane more fluid. This is called **capacitation**, and occurs several hours after sperm are deposited.

2. The acrosome releases protease enzymes which digest cells forming the corona radiate surrounding the oocyte allowing the sperm head to contact the zona pellucida. Now acrosin (another protease) hydrolyses the zona pellucida, allowing the head to enter the oocyte. This is called the **acrosome reaction**.

Key Terms

Capacitation: changes in the sperm membranes that increase its fluidity and allow the acrosome reaction to occur.

Acrosome reaction: acrosome enzymes digest the zona pellucida allowing the sperm and oocyte cell membranes to fuse.

Cortical reaction: occurs when the cortical granule membranes fuse with the oocyte cell membrane. The zona pellucida is converted into a fertilisation membrane.

3. The cell membranes of the sperm and oocyte fuse, the male nucleus can begin to enter the oocyte cytoplasm. This triggers the **cortical reaction**, where cortical granule membranes fuse with the oocyte cell membrane causing it to expand and harden forming the fertilisation membrane which prevents polyspermy –the further entry of sperm.

4. Meanwhile the second meiotic division completes, and the ovum is formed containing the female nucleus and a second polar body.

5. Fertilisation is the sequence of events from the point when the sperm and oocyte make contact until the male and female chromosomes join on the mitotic equator. The first mitotic division produces two cells, and the resulting cell is referred to as an embryo.

Implantation

The embryo continues to divide by mitosis as it passes down the oviduct, forming a ball of cells called a morula by day 3, by a process called cleavage. By day 7 a hollow ball of cells called a blastocyst forms which has an outer layer of cells referred to as a **trophoblast** which develops protrusions called trophoblastic villi. The endometrium thickens allowing **implantation** of the blastocyst to occur by day 8–10.

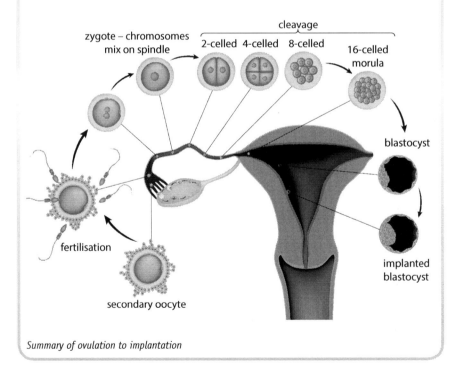

Summary of ovulation to implantation

The placenta

The chorion develops from the trophoblast, forming the larger chorionic villi which acquire capillaries and embed into the endometrium. They are supplied by the developing umbilical artery and vein, and form the placenta.

---

quicꜰɪʀe

⑤ Suggest why internal fertilisation is a necessary adaptation for terrestrial life.

quicꜰɪʀe

⑥ What is the function of the protease enzymes contained in the acrosome?

quicꜰɪʀe

⑦ What is the function of the cortical reaction?

Key Terms

Trophoblast: cells forming the outer layer of the blastocyst.

Implantation: the sinking of the blastocyst into the endometrium.

⑧ State two functions of the placenta.

⑨ State two functions of the amniotic fluid.

As the embryo develops into a foetus (where distinct organs are visible from about 10 weeks of pregnancy) the placenta takes on a number of roles:

- Allows exchange of oxygen, carbon dioxide, nutrients and waste between mother's and foetus's blood. The blood is never in direct contact, but is separated by just a few mm, and a counter current blood flow ensures that concentration gradients are maintained across the entire length of the placenta.
- Produces hormones to support pregnancy.
- Acts as a physical barrier between the two circulations. This is important as maternal blood pressure is much higher and would rupture delicate capillaries in the foetus, and it separates the maternal immune system from the foetus preventing an immune response.
- Antibodies can cross the placenta giving the foetus some passive immunity to diseases. However, some microorganisms can cross the placenta, e.g. *Rubella* virus, and many drugs, e.g. nicotine and heroin.

Blood leaves the foetus through the umbilical artery carrying waste including carbon dioxide to the placenta. The umbilical vein carries blood rich in nutrients, e.g. glucose and amino acids, back towards the foetus.

Pregnancy

Pregnancy lasts for around 39 weeks, is divided into three trimesters and runs from the first day of the last period until birth. As the foetus develops it is enclosed by a membrane called the amnion, which produces amniotic fluid by the fifth week. Amniotic fluid has a number of important functions. It:

- Acts as a shock absorber protecting the developing foetus.
- Helps to maintain the foetus's body temperature.
- Provides lubrication.
- Allows movement.

Hormones and birth

Human chorionic gonadotrophin (hCG) is secreted by the blastocyst, and later the chorion. hCG is responsible for maintaining the corpus luteum which secretes progesterone (maintains endometrium) up to about 16 weeks when progesterone is produced directly by the placenta. During pregnancy, progesterone also inhibits oxytocin preventing contraction of the myometrium, and oestrogen stimulates growth of the uterus and mammary glands.

To initiate birth, oxytocin is secreted by the posterior pituitary gland resulting in contractions of the myometrium. The contractions in turn cause an increased secretion of oxytocin and so on (positive feedback) causing more frequent and stronger contractions. Prolactin is secreted by the anterior pituitary gland causing milk to be produced, and is expelled from the nipples by contraction of muscles around the milk ducts brought about by oxytocin.

4.2 Sexual reproduction in plants

Flower structure

Flowering plants, or Angiosperms use flowers as their reproductive structures. The male gametes are contained within pollen which is produced within the anthers. The ovule(s) contains an embryo sac with one female gamete inside. To promote cross-pollination, the male and female parts of most flowers develop at different times. The sepals, which are usually green, protect the flower in the bud. The petals range from being absent, to small and green, to large and brightly coloured. The male parts of the flowers or stamens, consist of a filament supporting the anther which contains four pollen sacs. The female parts of the flower or carpels are found at the centre of the flower, and contain the ovary where the ovules develop. The stigma is the receptive surface which collects pollen during **pollination**.

quickfire
⑩ Describe two differences between the structure of insect- and wind-pollinated flowers.

quickfire
⑪ Name the three parts that make up the carpel.

Insect-pollinated flower

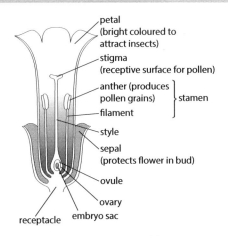

petal
(bright coloured to attract insects)
stigma
(receptive surface for pollen)
anther (produces pollen grains) } stamen
filament
style
sepal
(protects flower in bud)
ovule
ovary
receptacle   embryo sac

Diagram of a generalised insect-pollinated flower

- Large colourful petals, scent and nectar to attract pollinators such as insects.
- Anthers within the flower which transfer pollen to insects when they feed on nectar.
- Stigma within the flower to collect pollen from insect when it feeds on nectar.
- Small quantities of sticky sculptured pollen to stick to insect.

Wind-pollinated flower

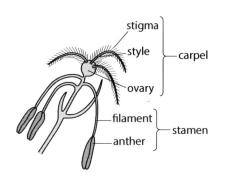

stigma
style
carpel
ovary
filament
stamen
anther

Diagram of a generalised wind-pollinated flower

- Small, green and inconspicuous, no scent, petals usually absent.
- Anthers hanging outside the flower so wind can blow pollen away.
- Large feathery stigmas providing a large surface area to catch pollen grains.
- Large quantities of small, smooth, light pollen to be carried by wind.

quickfire
⑫ What is the selective advantage of wind-pollinated plants producing light pollen?

> **Key Term**
>
> **Pollination**: the transfer of pollen from the anther of one flower to the mature stigma of another flower of the same species.

Development of gametes

Male gamete development

A pollen sac contains many pollen mother cells, each of which divides by meiosis to produce a tetrad. The surrounding nutritive layer called the tapetum provides nutrients to the developing pollen grains. The pollen cell wall is tough and resistant to desiccation. The haploid nucleus within the pollen grain undergoes mitosis to produce a generative nucleus and a pollen tube nucleus. The generative nucleus then undergoes further mitotic division producing two male nuclei. As the outer layers of the anther mature and dry out, the outer walls curl away exposing the pollen grains – a process known as dehiscence.

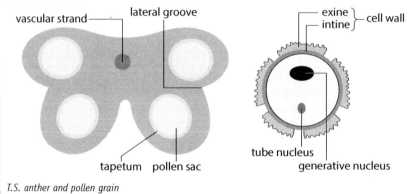

T.S. anther and pollen grain

Female gamete development

The ovules contain a megaspore mother cell which undergoes meiosis producing four haploid cells, only one of which develops further. It produces eight haploid cells following three mitotic divisions. Two of these cells fuse to produce a diploid polar nucleus, leaving six haploid cells: 3 antipodals, 2 synergids and 1 oosphere, all contained within the embryo sac which is surrounded by the integuments.

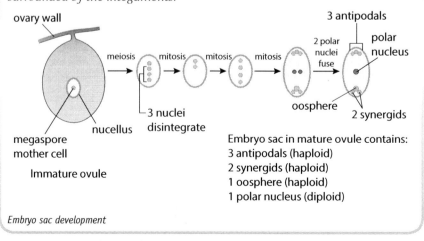

Embryo sac development

Key Term

Dehiscence: the opening of the anther, releasing pollen grains.

Grade boost

You could be asked to calculate the magnification of a drawing or the size of the original specimen. Remember your formula!

Pointer

Revisit your notes on mitosis and meiosis from AS.

Self and cross-pollination

Pollination is the transfer of pollen from the anther of one flower to the mature stigma of another flower of the same species. When pollination occurs between the anther and stigma in the *same* flower, or different flower on the *same* plant it is referred to as self-pollination. When pollen is transferred to another flower on a different plant of the same species, it is cross-pollination.

Self-pollination leads to self-fertilisation which results in inbreeding and so genetic variation is greatly reduced, as it only occurs as a result of mutation, independent assortment and crossing over. There is an increased risk of harmful recessive alleles coming together, but inbreeding does preserve successful genomes. Cross-pollination combines the genetic material from two different individuals so results in greater variation, and is referred to as outbreeding. Here the chance of harmful recessive alleles coming together is reduced, and allows for greater genetic variation potentially leading to evolution of species over time.

Ensuring cross-pollination

To ensure cross-pollination, plant species have evolved a number of methods.

- Stamen and stigma ripen at different times. When the stamens ripen first it is referred to as **protandry**.
- The anther is located below the stigma, reducing the risk of displaced pollen falling onto it.
- Some plants have separate male and female flowers, e.g. maize, or separate male and female plants, e.g. holly.
- Some plants show genetic incompatibility, e.g. red clover, where pollen cannot germinate on the stigma of the same plant.

Double fertilisation

Fertilisation is the process where the male gamete fuses with the female gamete, producing a diploid zygote. When a pollen grain lands on the mature stigma of another plant of the same species, (or same plant in the case of self-pollination) it germinates producing a pollen tube. The growth of the tube is controlled by the pollen tube nucleus, which also produces hydrolases, e.g. cellulases and proteases which digest a path through the style towards the micropyle and into the embryo sac guided by chemical attractants, e.g. GABA. The tube nucleus then disintegrates and the two male gametes enter the ovule. One male nucleus fuses with the haploid female nucleus, the oosphere, to form the zygote. The second male nucleus fuses with the diploid polar nucleus to form a triploid nucleus which develops into the endosperm which will provide nutrition for the developing embryo plant. This is referred to as double fertilisation.

Grade boost

A common mistake is to view self-pollination as a form of asexual reproduction. As pollen and ovules are produced by meiosis, some variation exists so the offspring are not genetically identical. Variation is however greatly reduced.

Key Terms

Protandry: stamens ripen before the stigmas.

Fertilisation: fusion of the male gamete with the female gamete, producing a diploid zygote.

Grade boost

Two fertilisations occur: one with the oosphere and the other with the diploid nucleus, hence the term double fertilisation.

quickfire

⑬ How is the endosperm formed?

quickfire

⑭ What is the function of the endosperm?

Structure of fruit and seed

Following fertilisation, the ovary wall becomes the fruit, whilst the ovule becomes the seed. Broad beans (*Vicia faba*) are dicotyledonous plants and as such have two seed leaves or cotyledons which absorb the food store or endosperm. The radicle forms the root, and the plumule forms the shoot. In monocotyledonous plants, e.g. maize, there is only one cotyledon. As the testa and ovary wall fuse, maize is actually a one seeded fruit! Seeds can remain dormant for many years.

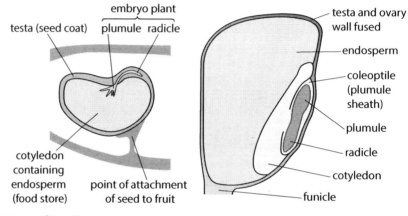

Diagram of broad bean seed   *Diagram of maize*

quickfire

⑮ Complete the following sentences: The diploid zygote divides by _____ to form the embryonic plant.

The _____ (the food reserve for the developing embryo) develops from the endosperm nucleus.

The _____ become the testa (seed coat) and the micropyle remains.

The fertilised ovule becomes the _____.

The wall of the fertilised ovary becomes the _____.

quickfire

⑯ What is the difference between pollination and fertilisation?

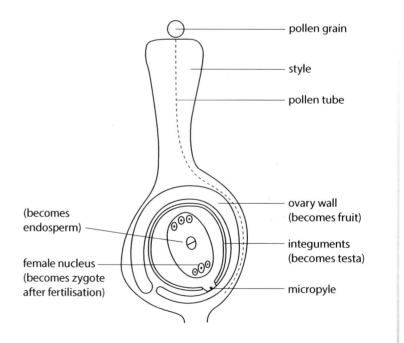

Growth of the pollen tube, and fate of the ovule post-fertilisation

Seed dispersal

Different plants have different methods of dispersing seeds. Seed dispersal is important, as it allows seedlings to germinate away from the parent plant and so reduces competition for resources. Some are carried by:

- Wind, e.g. dandelion seeds.
- Water, e.g. coconuts.
- Animals, attached to their fur, e.g. burdock.
- Animals, eat the fruits and egest the seeds away from the parent plant, e.g. cherries. The digestive system weakens the testa allowing germination to occur, and supplies its own fertiliser – faeces.

Germination in the broad bean

- Water is absorbed by the seed, causing the tissues to swell and mobilises the enzymes.
- The testa (seed coat) ruptures, the radicle pushes through first downwards, followed by the plumule upwards.
- Amylase enzyme hydrolyses starch into maltose which is transported to the growing points of the plant to be used in respiration.
- During germination the cotyledons remain below ground.
- The plumule is bent over in the shape of a hook to prevent damage to the tip by soil abrasion.
- When the plumule emerges from the soil it unfurls and begins to produce glucose by photosynthesising as the food reserves in the cotyledons have been now been depleted.

Requirements:

- Optimum temperature for enzyme action.
- Water for the mobilisation of enzymes and transport of products to growing points.
- Oxygen for aerobic respiration producing ATP for cellular processes such as protein synthesis.

As a seed **germinates** the dry mass of the cotyledons decreases as food reserves are used up fuelling growth of the embryo. The overall mass of the seed decreases initially until the plumule can begin photosynthesising.

Key Term

Germination: the biochemical and physiological processes through which a seed becomes a photosynthesising plant.

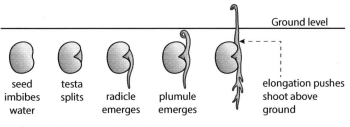

| seed imbibes water | testa splits | radicle emerges | plumule emerges | | elongation pushes shoot above ground |

Diagram of germination sequence in the broad bean

quickpire

⑰ Explain three requirements for germination.

quickpire

⑱ What is the importance of establishing a root system before a shoot system?

Effect of gibberellin

Gibberellic acid (GA) is a plant growth regulator which diffuses into the aleurone layer surrounding the endosperm switching on genes involved in transcription and translation, resulting in the production of amylases and proteases.

The amino acids produced by the hydrolysis of proteins are used to synthesise amylases which in turn hydrolyse stored starch into maltose and glucose for respiration by cells in the radicle and plumule.

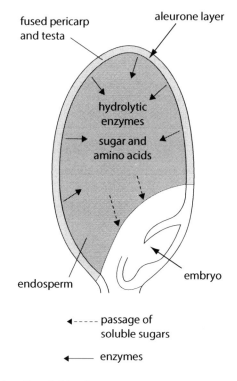

Maize fruit showing effect of gibberellic acid

4.3 Inheritance

Genes and alleles

A **gene** is a sequence of DNA on a chromosome normally coding for a specific polypeptide, which occupies a specific position or locus. Genes normally exist as two or more **alleles**, for example Rhesus blood group (positive or negative) or ABO blood group (three alleles I^A, I^B, I^O). The genotype of an organism is the genetic make-up, i.e. the actual alleles it possesses, whereas the phenotype represents the characteristics of the organism, which is the product of its genotype and the environment. When both alleles are the same, the organism is **homozygous** for that gene, e.g. RR or rr. When both alleles are different, the organism is **heterozygous** for that gene, e.g. Rr.

Performing genetic crosses

There are a few simple rules to make sure you present a genetic cross between two organisms in the correct way:

1. Choose a single letter to represent each characteristic, e.g. R.
2. Use upper case letters to represent **dominant** features (R), lower case for **recessive** (r), and state what they represent.
3. Clearly label PARENTS and circle GAMETES.
4. Use a matrix called a Punnett square to calculate crossing.
6. State the phenotype and ratios of offspring. First generation is represented by F_1, second by using F_2.

Monohybrid inheritance

Monohybrid inheritance involves the inheritance of a single gene. Gregor Mendel conducted a number of experiments with pea plants as they were easy to grow, showed clear differences in phenotypes, e.g. tall and dwarf plants, those with purple or white flowers, yellow or green seeds, and produced large numbers of seeds making the results reliable.

In one of Mendel's first experiments, he crossed peas with purple flowers with those with white flowers. He noticed that in the first generation (F_1) they were all purple, but when he self-pollinated the F_1 generation, the white flowers reappeared in a ratio of 3 purple to 1 white.

P is the allele for purple flowers, p for white.

Purple flowers v White flowers

(parents)	PP	pp
(gametes)	(P)	(p)
F_1	all are purple, Pp	

Grade boost

Learn your Key Terms carefully!

quickfire

⑲ Define recessive allele.

F1 generation were then self-pollinated.

(parents)	P p	×	P p
(gametes)	P , p		P, p

	P	p
P	PP	Pp
p	Pp	pp

1 PP (purple) 2 Pp (purple) and 1 pp (white)

This led Mendel to formulate his first law of inheritance called the law of segregation, which stated that *'The characteristics of an organism are determined by factors (alleles) which occur in pairs. Only one of a pair of factors (alleles) can be present in a single gamete.'*

Test or back cross

A test cross is performed to show if a dominant characteristic is determined by one or two dominant alleles, i.e. PP or Pp, and involves crossing the organism with the homozygous recessive. In the example above this would involve crossing a purple flower pea plant (either PP or Pp) with a white plant pp. If the F_1 generation are all purple then the purple plant was pure-bred or homozygous, but if there were 1 purple plant and 1 white plant then the plant used was not pure-bred, i.e. was the heterozygote Pp.

Co-dominance

In co-dominance both alleles involved are dominant and therefore both are expressed equally. An example of this is found in the ABO blood group where A and B are co-dominant. When showing co-dominance it is easier to use a letter to represent the gene, e.g. I and use superscripts to show the alleles as you have to use different letters.

$I^A I^A$ blood group is A

$I^B I^B$ group is B

$I^A I^B$ the group is AB

In snapdragons two flower colours exist: purple and white. When both alleles are present, the flowers appear pink – the phenotype is an intermediate rather than both alleles being expressed. This is called incomplete dominance.

$C^P C^P$	purple
$C^W C^W$	white
$C^P C^W$	pink

≫ Pointer

With co-dominance both alleles are expressed equally so both characteristics are seen. With incomplete dominance, an intermediate phenotype results.

Dihybrid inheritance

Mendel carried out experiments with pea plants involving two different characteristics at the same time, e.g. plants that produced yellow or green seeds AND wrinkled or round seeds. The simultaneous inheritance of two unlinked genes (genes on different chromosomes) is called dihybrid inheritance. Mendel noticed that the colour of the seed was inherited independently from the seed texture (wrinkled or round) and led to the second law called the law of independent assortment which stated that *'Each member of an allelic pair may combine randomly with either of another pair'.*

e.g. Pure bred (homozygous) yellow wrinkled peas were crossed with green smooth peas. In the F1 generation, all peas were yellow and wrinkled.

Key   allele for yellow colour Y   allele for green colour y

allele for round peas R   allele for wrinkled peas r

Parental phenotype

Yellow round seeds × Green wrinkled seeds

Parental genotype   YYRR   yyrr

Gametes   (YYRR)   (yyrr)

F1 genotype   YyRr

F1 phenotype   yellow round peas

F1 generation were self pollinated.

Phenotype   yellow round peas   yellow round peas

Genotype   YyRr   YyRr

Gametes

	(YR)	(Yr)	(yR)	(yr)
(YR)	YYRR yellow round	YYRr yellow round	YyRR yellow round	YyRr yellow round
(Yr)	YYRr yellow round	YYrr yellow wrinkled	YyRr yellow round	Yyrr yellow wrinkled
(yR)	YyRR yellow round	YyRr yellow round	yyRR green round	yyRr green round
(yr)	YyRr yellow round	Yyrr yellow wrinkled	yyRr green round	yyrr green wrinkled

Phenotype ratio

9 yellow round : 3 yellow wrinkled : 3 green round : 1 green wrinkled

Dihybrid inheritance

Mendel's actual data showed that 315 were round, yellow, 101 wrinkled, yellow, 108 round, green and 32 were wrinkled, green. Whilst this is not exactly 9:3:3:1 it is pretty close!

4.3

A yellow round pea (homozygous for colour but heterozygous for texture) was crossed with a homozygous green round pea. What are the genotypes and phenotypes of the offspring?

Autosomal linkage

This occurs when two different genes are found on the same chromosome and therefore cannot segregate independently (because they are on the same chromosome). This applies to autosomes, chromosomes other than the sex chromosomes. Using the same heterozygote yellow round example on the previous page (YyRr) only one homologous pair is needed to accommodate all four alleles, and the result is fewer possible combinations of gametes, YR and yr ONLY. Crossing over *could* occur if the two genes are not too close together, but this is a random event, so does not always occur on every chromosome pair. If crossing over *does* occur, the resulting gametes would be Yr and yR. Mendel's second law does NOT apply because the four possible gametes are not made in equal proportions.

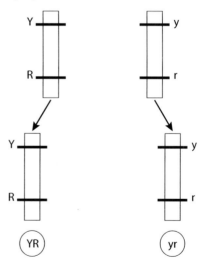

Possible types of gametes

Autosomal linkage

» **Pointer**

Don't get confused between sex linkage and autosomal linkage.

quickfire

⑳ What is meant by autosomal linkage?

Using statistics and probability

Are girls taller than boys? Look at the two graphs below.

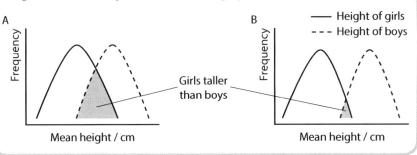

Significance

For us to say that boys are significantly taller than girls, this must be true in >95% (19 in 20) of cases, and therefore not the case in <5%. We can assign a degree of certainty indicating how sure we are using probability, p, where p = 1 is 100% and p = 0 is 0%. Therefore, to meet our criteria of significance, boys must be taller than girls in at least 95% of cases, i.e. p = 0.95. The probability that boys are *not* taller than girls, i.e. any differences seen were due to chance, must be in less than 5% of cases, i.e. < p = 0.05. This is referred to as the critical value.

Grade boost

Remember the critical value p must be < 0.05

Null hypothesis

In our example above, our actual hypothesis (called the alternative hypothesis) is that boys are taller than girls. The null hypothesis would be that there is no difference between the mean height of boys and girls. You must write both down, and using the statistical test, say whether you accept or reject the null hypothesis and why.

Pointer

You need to be able to state a null hypothesis for a given experiment.

Critical value

Using different statistical tests, you will calculate a value (calculated value) that must exceed a *critical value* (at p = 0.05) for that particular number of data sets and that particular test, if your result is to be significant.

If your calculated value exceeds the critical value (at p = 0.05), in our example here there is a significant difference between the height of boys and girls.

In the case of a Chi squared test, if your calculated value exceeds the critical value (at p = 0.05), there would be a significant difference between the observed and expected results, so you can reject your null hypothesis, as any differences seen were not due to chance at p = 0.05. (This means that Mendel's laws DO NOT apply.)

If your calculated value is less than the critical value (at p = 0.05), there is NOT a significant difference between observed and expected results, you must accept your null hypothesis, as any differences seen were due to chance at p = 0.05. (Mendel's laws apply.)

Using the chi-squared statistical test

In guinea pigs the allele for black coat (B) is dominant to the allele for albino (b) and the allele for rough coat (R) is dominant to the allele for smooth coat (r). A heterozygous black smooth coated guinea pig is mated with an albino guinea pig which is heterozygous for rough coat. In the first generation the offspring had the following phenotypes: 27 black rough coat; 22 black smooth coat; 28 albino rough coat; 23 albino smooth coat. Mendel's dihybrid inheritance would predict a 1:1:1:1 ratio or 25:25:25:25.

We can use χ^2 to find out if there is a significant difference between the observed and expected numbers of offspring of the different phenotypes.

Null hypothesis: there is no difference between observed and expected numbers.

Chi squared is calculated from

>> **Pointer**

Σ means sum of.

$$\chi^2 = \Sigma \frac{(O-E)^2}{E}$$

By using a table the formula is broken down into manageable steps:

1. Enter the observed numbers (O) into the table.

2. Calculate the expected numbers (E) from the total observed, i.e. 100 and the expected ratios, e.g. 1 in 4 will be black rough = 0.25 × 100 = 25.

3. Take observed value away from its corresponding expected value, then square that value.

4. Divide *each* (O – E)^2 value by its expected value, e.g. $\frac{4}{25}$ = 0.16.

5. Then add all these up to get χ^2.

Category	O	E	O–E	(O–E)^2	$\frac{(O-E)^2}{E}$
black rough coat	27	25	2	4	0.16
black smooth coat	22	25	3	9	0.36
albino rough coat	28	25	3	9	0.36
albino smooth coat	23	25	2	4	0.16
	Σ 100				Σ = 1.04

χ^2 = 1.04

Then use a chi-squared table to see if the calculated value exceeds the critical (table) value.

>> **Pointer**

The degrees of freedom = the number of classes –1.

>> **Pointer**

To determine the critical value, use the p = 0.05 column.

Degrees of freedom	P = 0.10	P = 0.05	P = 0.02
1	2.71	3.84	5.41
2	4.61	5.99	7.82
3	6.25	**7.82**	9.84
4	7.78	9.49	11.67
5	9.24	11.07	13.39

Table (critical value at p = 0.05 = **7.82** (degrees of freedom = n–1 i.e. 3).

Because the calculated value is less than the critical value at p=0.05 (1.04 < 7.82); *we can accept the null hypothesis, any difference seen was due to chance.* The converse would be true if the χ^2 exceeded the critical value.

Sex determination

Humans have 46 chromosomes arranged into 23 pairs; 22 pairs 'match', i.e. contain the same genes (but not necessarily same alleles) so are said to be homologous, and are the *autosomes*. The 23rd pair or sex chromosomes match in females (XX) but are different in males (XY), and these determine sex. In humans, there is a 50:50 chance that any child will be male.

Sex linkage

Sex linkage is when a gene is carried by a sex chromosome so that a characteristic it encodes is seen predominately in one sex. As males contain one X and one Y chromosome, the Y must come from his father, the X from his mother. The Y chromosome is smaller than X, and there are some genes that are only carried on the X chromosome so males only receive one copy.

quickfire

㉑ Define sex linkage.

Haemophilia

The gene for clotting factor VIII, needed to clot the blood following injury, is only carried on the X chromosome. Haemophilia is a recessive sex-linked, X chromosome disorder, which is more likely to occur in males rather than females, because females have two X chromosomes while males have only one. If a male receives the recessive haemophiliac allele (which codes for the altered faulty factor VIII causing haemophilia), he will suffer from the disease, as he lacks a second X chromosome that might carry a normal allele. Females are almost exclusively carriers of the disorder as they may inherit the defective allele from their mother or father, or as a result of a new mutation. Only under rare circumstances do females actually have haemophilia as they would need to inherit the faulty allele from both parents (homozygous recessive).

Duchenne muscular dystrophy (DMD)

Like haemophilia, DMD is caused by an X-linked recessive allele, but involves the gene which codes for dystrophin, a component of a glycoprotein that stabilises the cell membranes of muscle fibres. The disease is characterised by progressive muscle loss and weakness. Like haemophilia, the defective allele is passed to children from their mothers (as boys receive the Y chromosome from their father).

X^D is normal dystrophin allele

X^d is allele coding for faulty dystrophin resulting in DMD.

The outcome of a carrier female having children with a normal male is shown below.

	X^D	Y
X^D	$X^D X^D$	$X^D Y$
X^d	$X^d X^D$	$X^d Y$

1 $X^D X^D$ female normal

1 $X^d X^D$ female carrier

1 $X^D Y$ male normal

1 $X^d Y$ male DMD

quickfire

㉒ Using the symbol X^H to represent the allele for normal blood clotting factor VIII, and X^h to represent the allele for altered factor VIII that produces haemophilia, what is the probability of a carrier mother having a male child with haemophilia?

quickfire

㉓ Using the pedigree diagram, explain the phenotypes of parents 1 and 2.

Pedigree diagrams

These are family trees that show the instances of a particular inherited condition within a family. The diagram shows the inheritance of haemophilia within a family.

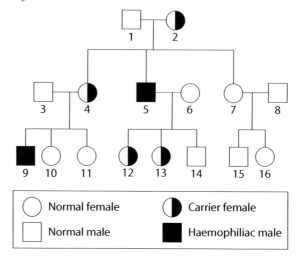

Pedigree diagram

The diagram shows us that as haemophilia only occurs in males in this family, it is likely to be sex-linked, and as it is inherited through the mother it is likely to be carried on the X chromosome.

Mutations

A mutation is a change in the amount, arrangement or structure of the DNA in an organism. There are two types:

1. Gene or point mutations, e.g. sickle cell disease, caused during DNA replication.

2. Chromosome mutations cause changes in the structure or number of whole chromosomes in cells. They result in an entire chromosome being added or lost (called aneuploidy, e.g. trisomy 21 or Down's syndrome), or when one chromosome breaks off and attaches to another chromosome called translocation, e.g. translocation Down's.

Mutations can be spontaneous and are random events, occurring with equal probability anywhere in the genome of diploid organisms. They may contribute to evolution if it provides advantages which can be selected for. Generally, mutation rates are very low, but are increased by ionising radiation, X-rays, polycyclic hydrocarbons in tobacco smoke, and some chemicals, e.g. benzene. Mutations are more common in organisms with short life cycles, e.g. in bacteria where the life cycle is as short as 20 minutes. Most mutations occur during crossing over in prophase I and non-disjunction in anaphase I and anaphase II of meiosis. Mutations that occur in somatic (body) cells are non-heritable.

 Grade boost

Mutations do occur randomly, but they do occur with a set frequency which varies between species. Mutagens increase the rate at which mutations occur.

Gene or point mutations

Point mutations occur as a result of:

- Addition or subtraction: where a base is added/deleted. Both result in a frame shift, where the reading frame moves one place and usually results in a non-functional protein as the amino acid sequence changes significantly.

- Substitution: where one base is 'swapped' for another, which may result in a change of codon, and hence amino acid.

Many point mutations have no effect because the change:

- Is silent, i.e. the base changes but the amino acid for which the codon codes does not.

- May be in a non-coding region or intron.

- May be in a recessive allele and so is not expressed.

- Alters an amino acid but may not result in a change to the functioning of the protein.

An example of a point mutation is sickle cell disease, which is common in Afro-Caribbean, Middle Eastern, Eastern Mediterranean and Asian populations that evolved in malarial habitats. In sickle cell, a mutation involves a substitution of adenine for thymine which does result in the change of one amino acid to valine, causing red cells to deform and block capillaries under low partial pressures of oxygen. The alleles for normal and affected haemoglobin are co-dominant, so people who have one copy of the faulty allele do have symptoms, but not as severe as sufferers, but show increased resistance to malaria which is an example of heterozygote advantage.

Chromosome mutation

Down's syndrome is a chromosomal disorder occurring in around 1 in 800 births, and results when a person inherits all or part of an extra copy of chromosome 21. This occurs most commonly following non-disjunction of chromosome 21 during anaphase 1 or 2 of meiosis, where both copies of chromosome 21 enter the gamete. When this is then fertilised by a normal gamete, three copies of chromosome 21 result (trisomy 21). The oocyte containing no copies of chromosome 21 fails to develop further. The risk of Down's syndrome is related to the mother's age:

> 18 yr old = 1 in 2100 births
>
> 30 yr old = 1 in 1000 births
>
> 40 yr old = 1 in 100 births

There is no treatment, but it can be diagnosed by prenatal tests, e.g. amniocentesis.

Grade boost

Not all mutations result in a change to the protein produced. Some may be silent, in recessive alleles or introns, or the resulting amino acid change does not affect the protein's function.

Pointer

Trisomy means three copies of a chromosome.

Cancer

A mutagen that causes cancer is known as a carcinogen. Cancer results from mutations in protooncogenes producing oncogenes that cause the rate of cell division to increase either by resulting in the production of excess growth factor, or by mutated receptor proteins that do not require growth factor to initiate cell division.

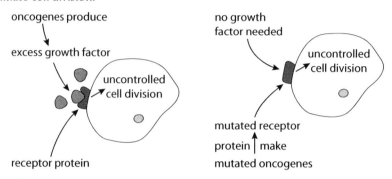

Role of mutated oncogenes (protooncogenes)

Tumour suppressor genes, or anti-oncogenes slow cell division. A mutated suppressor gene also causes the rate of cell division to increase because the gene is inactivated.

quickfire

㉕ What is the word for mutagens that cause cancer?

Control of gene expression – Epigenetic modifications

Variation within a population is due to both allelic variations and the environment. **Epigenetic** modifications occur as a result of environmental conditions, causing changes to how genes are transcribed. Visible differences seen in identical twins are largely due to epigenetic modifications. Epigenetic modifications are caused by:

1. DNA methylation, which involves the addition of a methyl or hydroxymethyl group to cytosine, reducing transcription of the gene.

2. Histone modification post-translation, resulting in changes to the way the histones interact with DNA by causing a looser arrangement of the nucleosomes. This causes increased transcription because RNA polymerase and other transcription factors have easier access to the DNA. Histone modification can involve the addition of an acetyl or methyl group to lysine, and arginine or phosphate groups to serine and threonine.

As stem cells differentiate, epigenetic changes result in the production of different proteins, e.g. melanin within skin cells, and amylase within pancreatic cells. Some epigenetic modifications can be heritable if changes occur in gametes which is referred to as genomic imprinting, and whole chromosomes can be inactivated, e.g. X chromosome inactivation which results in the patchwork of tortoiseshell cats.

Key Term

Epigenetics: is the control of gene expression by modifying DNA or histones, without modification of the base sequence.

Link Post-translational modification was covered in AS.

4.4 Variation and evolution

Types of variation

Organisms show variation in their phenotypes due to having different genotypes, having the same genotype but different epigenetic modifications, or by being exposed to different environments. Variation which is heritable, i.e. can be passed onto offspring, arises from:

- Gene (point) mutations.
- Crossing over during prophase I of meiosis.
- Independent assortment during metaphase I and II of meiosis.
- Random mating, i.e. that any organism can mate with another.
- Random fusion of gametes, i.e. the fertilisation of any male gamete with any female gamete.
- Environmental factors leading to epigenetic modifications.
- Environmental factors can also lead to non-heritable variation within a population, e.g. diet.

There are two types of variation:

1. Continuous variation, e.g. height.
   - Range of phenotypes seen.
   - Controlled by many genes (polygenic).
   - Follows a 'normal' distribution.
   - Environmental factors have a major influence, e.g. diet on weight.

2. Discontinuous variation, e.g. blood group.
   - Characteristics fit into distinct groups.
   - There are no intermediates.
   - Usually controlled by one gene with two or more alleles (monogenic).
   - Environmental factors have little influence, e.g. diet has no effect on blood group.

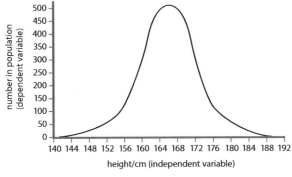

Normal distribution curve

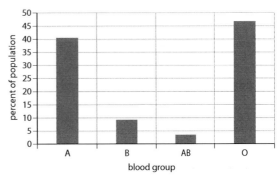

ABO blood group distribution

Statistical significance of continuous variables using the *t*-test

The mean is a measure of central tendency. It is most commonly calculated as the arithmetic mean which is the sum of the values divided by the number of values. This should not be confused with the mode, which is the most common value.

The standard deviation (SD) is a measure of the variation in the data either side of the mean. If the data is normally distributed then 95.4% of the data will lie within 2× standard deviations either side of the mean.

Statistically, if two data sets are significantly different, then no data will overlap within mean + 2 SD.

To compare means of data values of two populations, the *t*-test uses the formula.

$$t = \frac{\bar{x}_1 - \bar{x}_2}{\sqrt{\dfrac{s_1^2}{n_1} + \dfrac{s_2^2}{n_2}}}$$

where $\bar{x}$ = mean of observations

n = number of observations (sample size)

s = standard deviation

(The subscripts 1 and 2 refer to samples 1 and 2 respectively.)

Note: The two samples must have the same number of observations.

1. Work out the means of each sample.

2. Then subtract one from the other.

3. Then work out the standard deviation for each sample by subtracting the observed value from the mean for EACH observation.

4. Then square EACH value and THEN add them up.

5. Divide this value by the number of observations minus 1 and square root your answer.

$$\sqrt{\frac{\Sigma(x - \bar{x})^2}{n - 1}}$$

6. Work out the standard deviation for the other sample.

7. Square your deviation and divide by the number of observations in that sample, do the same for the other sample, add together and square root your answer.

8. Finally, divide the differences in the means by

$$\sqrt{\frac{s_1^2}{n_1} + \frac{s_2^2}{n_2}}$$

where S_1 is standard deviation for sample 1, and S_2 sample 2.

Your calculated *t*-value must *exceed* the critical value in the table for 0.05 (5% probability) for the degrees of freedom (total number of observations −2) for you to be sure differences seen were not due to chance. See page 97 'Using statistics and probability'.

quickfire

㉖ Pea plants can be tall or short. What can you conclude about the type of variation they show?

quickfire

㉗ State four sources of variation.

Environmental factors

Environmental influences affect the way a genotype is expressed and result in different phenotypes, e.g. industrial melanism in the peppered moth *Biston betularia*. There are two forms of the peppered moth: speckled and dark (melanic). In polluted habitats where trees are covered with soot the dark form prevails, but in unpolluted habitats where lichens are found the speckled form is more common. In each instance the moth's colour camouflages it against the environment conferring a selective advantage, so it is more likely to survive, and reproduce transmitting advantageous alleles to the next generation and so the numbers increase within the population.

Competition

There are two types of competition:

- Intraspecific competition is where members of the *same* species vie for the same resource in an ecosystem, e.g. food, light, nutrients, availability of nesting sites.
- Interspecific competition is where individuals of *different* species vie for the same resource in an ecosystem, e.g. different plant species competing for water.

The different resources being competed for act as a **selection pressure**, therefore individuals with an advantage meaning they are *more* successful in gaining food and shelter are *more* likely to survive and pass those advantageous alleles onto the next generation. The increased chance of survival and reproduction of organisms with phenotypes suited to their environment is called **natural selection** and can lead to evolution.

Population genetics

The **gene pool** of a population is all of the alleles present in a population at any one time. Population genetics considers the relative proportions of the different alleles, or **allele frequencies** within the gene pool. If the environment is stable then allele frequencies also remain stable; however, environments do change and bring different selection pressures which favour some alleles over others, so their frequency increases.

Genetic drift and the founder effect

Genetic drift is the chance variations in the relative frequency of alleles in a population. This is due to *random sampling* (which alleles are inherited) and *chance* (that an individual may survive and breed). Genetic drift leads to changes in allele frequencies over time, and is most significant in small or isolated populations where a change will constitute a much larger proportion of the population because the population is small, and so it may be an important evolutionary process.

Key Terms

Selection pressure: an environmental factor that can alter the frequency of alleles in a population, when it is limiting.

Natural selection: the increased chance of survival and reproduction of organisms with phenotypes suited to their environment, enhancing the transfer of favourable alleles from one generation to the next.

quickfire

㉘ What is the effect of natural selection on allele frequency?

Key Terms

Gene pool: all the alleles present in a population at a given time.

Allele frequency: the frequency of an allele is its proportion, fraction or percentage of all the alleles of that gene in a gene pool.

Genetic drift: chance variations in allele frequencies in a population.

When a small number of individuals become isolated and start a new population, say by colonising a new island, the founder members of the new population are a small sample of the population from which they originated, (the **founder effect**) and may be subject to genetic drift. This was seen in the adaptive radiation seen in Darwin's finches on the Galapagos Islands.

The Hardy-Weinberg principle

The Hardy-Weinberg principle states that in ideal conditions the allele and genotype frequencies in a population are constant. Under ideal conditions:

- Organisms are diploid, have equal allele frequencies in both sexes, reproduce sexually, mating is random and generations don't overlap.
- The population size is very large, and there is no migration, mutation or selection.

The Hardy-Weinberg equation is $p^2 + 2pq + q^2 = 1$
Where p represents the frequency of the dominant allele, e.g. A
Where q represents the frequency of the recessive allele, e.g. a
Therefore $p + q = 1$

p^2 = frequency of homozygous dominant, e.g. AA
$2pq$ = frequency of heterozygous, e.g. Aa
q^2 = frequency of homozygous recessive, e.g. aa

It is therefore possible to use the equation to calculate allele and genotype frequencies from the instances of a disease or other phenotypic character.

Grade boost

Remember your p's and q's!
$p^2 + 2pq + q^2 = 1$ and $p + q = 1$.

Worked example

Cystic fibrosis is a recessive condition affecting around 1 in 2500 babies. Calculate the frequency of the recessive allele and the proportion of carriers in the population.

$q^2 = \dfrac{1}{2500} = 0.0004$

$q = \sqrt{0.0004} = 0.02$ so allele frequency (q) = 2%

as $p + q = 1$, $p = 1 - 0.02 = 0.98$

frequency of heterozygotes (Aa) i.e. 2pq,
so frequency = $2 \times 0.98 \times 0.02 = 0.0392 = 0.04$ (2 dp)

$0.04 \times 100 = 4\%$ or 1 in 25 are carriers.

4.4

Thalassaemia is a recessive genetic condition that results in the production of too little haemoglobin, resulting in anaemia, and is controlled by one gene with two alleles. It mainly affects people of Mediterranean, South Asian, Southeast Asian and Middle Eastern origin, resulting in 4.4 cases in every 10 000 live births worldwide annually. Calculate the proportion of carriers in the population in the same year.

Natural selection

Organisms overproduce offspring so that there is a large variation of phenotypes in the population. Changes to environmental conditions bring new selection pressures through competition/predation/disease which results in a change in the allele frequency.

Isolation and speciation

Speciation is the evolution of new **species** from existing ones.

There are two types:

1. **Allopatric speciation** – involves geographical isolation which reproductively isolates two sub-groups (demes) within a population of the same species, preventing gene flow between them. This is followed by exposure to different environmental conditions that favour different individuals within each deme. After thousands of generations exposed to the differing conditions, allele frequencies within the demes change as a result of different mutations. If the barrier is removed, the two populations have changed so much that they can no longer interbreed.

2. **Sympatric speciation** – involves populations living together becoming reproductively isolated by means other than a geographical barrier, e.g. behavioural isolation occurs in animals with elaborate courtship behaviours where members of a sub-species fail to attract the necessary response, e.g. sticklebacks. Other mechanisms include:

   - Seasonal (temporal) isolation where organisms are isolated due to reproductive cycles not coinciding and so are fertile at different times of the year. This is seen in frogs where each of four types has a different breeding season, e.g. wood frog, pickerel frog, tree frog and bullfrog.

   - Mechanical isolation as a result of incompatible genitalia.

   - Gametic isolation from the failure of pollen grains to germinate on stigma or sperm fail to survive in oviduct, e.g. fruit flies.

   - Hybrid inviability – embryo development may not occur.

Key Term

Species: a group of individuals with similar characteristics that can interbreed to produce fertile offspring.

≫ Pointer

Reproductive isolation can be pre-zygotic (prevents gametes from fusing to form a zygote), or post-zygotic (zygote is formed but developing organism is sterile).

Hybrid sterility

Hybrid sterility results when two closely related species interbreed but due to differences in chromosome structure or number, chromosomes fail to pair during prophase I of meiosis and so gametes do not form. The resulting offspring are sterile.

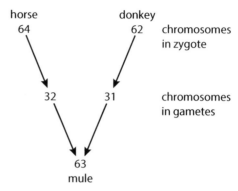

Hybrid sterility in mules

Darwinian evolution

Evolution is the process by which new species are formed from pre-existing ones over a long period of time. Darwin's observations of variation within a population led to the development of the theory of natural selection and Darwinian evolution in the 19th century. In short:

- Organisms overproduce offspring, so that there is a large variation of genotypes in population.

- Changes to environmental conditions bring new selection pressures through competition/predation/disease.

- Only those individuals with beneficial alleles have a selective advantage, e.g. sandy colour of the fennec fox that lives in deserts of North Africa and Asia, which helps provide camouflage, increasing their chances of survival.

- These individuals then reproduce more successfully than those without the beneficial alleles.

- Offspring are likely to inherit the beneficial alleles.

- Therefore the beneficial allele frequency increases within the gene pool.

4.5 Application of reproduction and genetics

Human Genome Project

The Human Genome Project began in 1990 and took ten years to complete, but the analysis of all the sequences obtained took much longer. The main aims of the project were to:

1. Identify all the genes in the human genome and identify their loci (positions on the chromosome).

2. Determine the sequence of the 3.6 billion bases present in the human genome and store in databases.

3. Consider the ethical, social and legal issues that arise from storing this information.

The project found:

- The number of genes present in the human genome is around 20,500.

- There are large numbers of repeating sequences called STRs (short tandem repeats).

Sanger sequencing

The project used a method of sequencing called 'Sanger sequencing' named after the scientist who invented it. It works by sequencing small fragments of DNA around 800 bases in length created by the use of **restriction enzymes**. DNA polymerase was then used to synthesise complementary strands using the **polymerase chain reaction**. Four reactions were carried out (one for adenine, thymine, cytosine and guanine), each containing complementary nucleotides marked with a radioactive marker, but a proportion of the nucleotides used in each reaction had been altered (stop nucleotides). When these were incorporated into the complementary strand, further synthesis was prevented.

Take the following sequence:

5' AGC**T**AGCCCGG**T**AGACC 3'

In the thymine reaction, it is a random event whether a normal thymine nucleotide is incorporated at base 4 or a stop nucleotide is incorporated, but over the course of many reactions, some DNA strands produced will have incorporated a stop nucleotide and some won't. The same is true for thymine at position 12. The result of the reaction will be some DNA fragments 4 bases long and some 12. When the results for all the reactions for each nucleotide are run out side by side on an agarose gel using **electrophoresis**, and the resulting gel exposed to x-ray film to detect the radioactive signal, the sequence can be determined by reading the banding pattern because

Key Terms

Restriction enzymes: bacterial enzymes that cut DNA at specific base sequences.

Polymerase chain reaction: a technique that produces a large number of copies of specific fragments of DNA, rapidly.

Electrophoresis: a technique that separates molecules according to size.

electrophoresis separates DNA fragments according to size. Working from the bottom up, the sequence order is A G C T, etc.

The Sanger method is very slow, taking days to accurately sequence a few thousand bases. With the introduction of Next Generation Sequencing (NGS), entire genomes can be sequenced in hours.

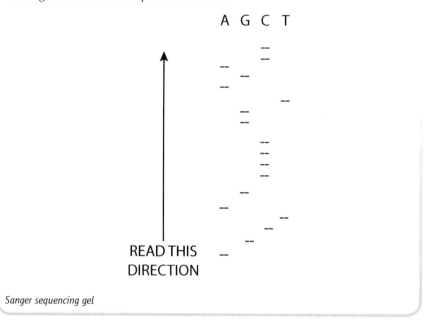

READ THIS DIRECTION

Sanger sequencing gel

quickfire

③① Explain how the electrophoresis method works.

The 100K Genome Project

Launched in 2012 using NGS, the project aims to sequence 100,000 genomes from healthy individuals and patients with medical conditions across the UK to establish any variance in their base sequences and identify if there is a genetic correlation. It is hoped that diseases will be better understood and new treatments can be found.

Moral and ethical concerns

Society has still to decide how the vast information collected through the Human Genome and 100K projects should be treated. Ethical concerns exist:

- If a patient has a genetic predisposition to a particular disease, should this information be passed to life or health insurance companies?
- If ancestral relationships are determined, this could be used to socially discriminate against people.
- If genetic diseases are identified, this has an implication for the parents and children of those diagnosed. If children are screened, when should they be told if they have a predisposition say for Alzheimer's disease?
- Could screening of embryos be extended from genetic diseases to desirable traits?
- How to ensure safe storage of patient data.

Mosquito genome project

Other organisms have had their genomes sequenced, which has allowed us to determine evolutionary relationships, e.g. how closely related we are to primates. The mosquito *Anopheles gambiae* which is responsible for transmitting malaria to around 200 million people annually, has also had its genome sequenced in attempts to tackle insecticide resistance in the vector. Malaria is discussed in more detail on page 126.

In 2015 gene-editing technology was used to produce a genetically modified mosquito that could produce antibodies to the *Plasmodium* parasite that it transmits. Whilst the mosquito won't be released into the wild, it is an exciting step forward in the control of malaria.

Other attempts to control malaria have focused on the parasite, *Plasmodium*. It too has developed resistance to many of the drugs used to treat it, e.g. atovaquone, larium, artimesinin, but it is hoped that the sequencing of its genome will allow for the development of new drugs.

Polymerase chain reaction (PCR)

Using the polymerase chain reaction (PCR) technique, a large number of copies of specific fragments of DNA may be made rapidly. From each strand of DNA it is possible to produce over a billion copies in a few hours.

PCR requires:

- A heat stable DNA polymerase isolated from the bacterium *Thermus aquaticus*, which lives in hot springs.
- Short single-stranded pieces of DNA called **primers** (6–25 bases long) which act as a start point for the DNA polymerase, and are complementary to the start point on the DNA strand of interest.
- Deoxyribonucleotides containing the four different bases.
- A buffer.

During the reaction, a thermocycler is used to rapidly change the temperature.

1. Step 1 – heat to 95°C to separate the DNA strands by breaking the hydrogen bonds between the two complementary DNA strands.
2. Step 2 – cool to 50–60°C to allow the primers to attach by complementary base pairing (annealing).
3. Step 3 – heat to 70°C to allow the DNA polymerase to join complementary nucleotides (extension).
4. Repeat 30–40 times.

PCR has some limitations:

- Any contamination is quickly amplified (copied).
- DNA polymerase can sometimes incorporate the incorrect nucleotide (about once every 9000 nucleotides).
- Only small fragments can be copied (up to a few thousand bases).
- The efficiency of the reaction decreases after about 20 cycles, as the concentrations of reagents reduce, and product builds up.

Key Term

Primer: a short single strand of DNA between 6 and 25 bases long that is complementary to the base sequence at one end of a singled-stranded DNA template, acting as a start point for DNA polymerase to attach.

» Pointer

30 cycles of PCR will produce 2^{30} copies of DNA – that is just over 1 billion!

quickfire

③② Why does the temperature need to be lowered to 50–60°C in PCR?

Grade boost

It is important that you know what happens at each of the three steps in PCR and the temperatures needed to accomplish it.

Genetic fingerprinting

A breakthrough came in the mid-1980s when Professor Alec Jeffreys at Leicester University was able to use the many variable regions of DNA which did not code for amino acids called **short tandem repeats (STR)** regions, to produce a genetic fingerprint. These regions were called microsatellites and there are thousands of them scattered throughout the chromosomes. The number of times that these regions are repeated gives individuality. PCR is then used to amplify specific microsatellite sequences from very small samples of DNA left at a crime scene.

An example of a STR is D7S280, a repeating sequence found on human chromosome 7. The DNA sequence of a representative allele of this locus is shown below. The tetrameric repeat sequence of D7S280 is 'gata'. Different alleles of this locus have from 6 to 15 tandem repeats of the 'gata' sequence, so the more times it repeats, the larger the fragment of DNA will be.

1   aatttttgta ttttttttag agacggggtt tcaccatgtt ggtcaggctg actatggagt

61   tattttaagg ttaatatata taaagggtat gatagaacac ttgtcatagt ttagaacgaa

121   ctaac**gatag atagatagat agatagatag atagatagat agatagatag atagata**gat

181   tgatagtttt tttttatctc actaaatagt ctatagtaaa catttaatta ccaatatttg

241   gtgcaattct gtcaatgagg ataaatgtgg aatcgttata attcttaaga atatatattc

301   cctctgagtt tttgatacct cagattttaa ggcc

In the example above, '**gata**' shown in bold repeats 13 times, so the size of the DNA fragment produced will be 13 × 4 bases = 52 bp. Currently ten different microsatellite sequences are used to build up a unique fingerprint in UK (13 in US). When these different-sized fragments are visualised by gel electrophoresis a unique banding pattern is created. To visualise DNA, ethidium bromide is often used as it intercalates with DNA (inserts between the base pairs) and fluoresces under ultraviolet light.

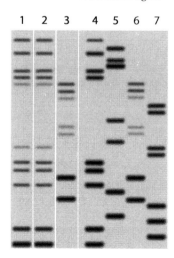

Genetic fingerprint © N. Roberts

Gel electrophoresis

Gel electrophoresis is a method of separating DNA fragments according to size.

The gel used is made from agarose (a polysaccharide which is the main constituent of agar) which contains pores in its matrix. DNA samples are loaded and a voltage is applied across the gel. DNA is attracted to the positive electrode due to its negative charge on the phosphate groups.

Smaller fragments migrate more easily through the pores in the gel and so travel further than large fragments in the same time. Size of fragments can be estimated if a sample of known DNA sized fragments (called a DNA ladder) is run at the same times as the samples.

DNA **probes** can also be used to find DNA sequences of interest within DNA fragments. Probes, which are portions of single stranded DNA incorporating a radioactive tracer (^{32}P) or fluorescent label, are designed to be complementary to part of the sequence of interest. When the probe is washed over the gel, it binds to the exposed complementary nucleotides by a process known as **DNA hybridisation**. The DNA fragment which contains the sequence of interest is then identified by its fluorescence or radioactive signal. To detect a radioactive signal, the DNA from the gel is transferred to a nylon membrane, and the membrane is then exposed to X-ray film, producing an autoradiograph.

Key Terms

Probe: short piece of DNA that is labelled with a fluorescent or radioactive marker, used to detect the presence of a specific base sequence in another piece of DNA, by complementary base pairing.

DNA hybridisation: single-stranded DNA molecules anneal to complementary DNA.

quickFire

㉝ Why does DNA migrate towards the positive electrode in electrophoresis?

Use of genetic fingerprinting – DNA profiling

DNA profiling is a non-invasive procedure requiring hair samples, or a mouth swab to collect enough DNA which can be further amplified by PCR. It has been used successfully to provide evidence in criminal cases, as well as a number of different situations:

- To provide forensic evidence to identify or rule out suspects, or to identify human remains.

- To prove paternity, or in rare cases maternity. Here the genetic fingerprint of the child is composed of elements of the fingerprints of both parents. It has also been used to identify siblings.

- In immigration applications where the right to remain in a country exists for a parent and their children.

- Phylogenetic studies where relatedness of species can be investigated to suggest evolutionary links.

DNA profiling cannot guarantee a match: at best a genetic fingerprint has a 1 in 1 billion chance that someone else could have the same profile, which still leaves some uncertainty. Ethical and legal concerns exist over the storage of DNA profiles by agencies such as the police, or health insurance providers, and the safe storage of personal data. DNA evidence in criminal cases is often relied upon too much to prove guilt, instead of supporting other evidence: a positive DNA sample from a crime scene may strongly indicate that a particular individual was present, not that they necessarily committed the crime.

Genetic engineering

Genetic engineering allows genes to be manipulated, altered or transferred from one organism or species to another, making a genetically modified organism (GMO). It has been used successfully to produce insulin by inserting the human insulin gene into bacteria, to produce disease resistant crops, and with some success to reduce the symptoms of some genetic diseases, e.g. Duchenne muscular dystrophy (DMD).

When the genetic material from two different species is combined, the result is **recombinant DNA**, and if donor DNA is inserted into another organism, the organism becomes **transgenic**. The main tools used in genetic engineering are restriction enzymes.

Recombinant DNA: DNA produced by combining DNA from two different species.

Transgenic: an organism that has been genetically modified by the addition of a gene or genes from another species.

Restriction enzymes

>> **Pointer**
Restriction enzymes are named after the bacteria they have been isolated from, e.g. *EcoR1* comes from *E.coli*.

Restriction enzymes (restriction endonucleases) are bacterial enzymes that cut up any foreign DNA which enters a cell. When a restriction enzyme makes a cut, the cut may be staggered and there are short, single-stranded fragments at either end. These overhangs are called 'sticky ends'. Sometimes these enzymes cut DNA between specific base sequences which the enzyme recognises and do not leave a 'sticky end'. These are called 'blunt cutters':

```
A │ A   G   C   T   T
                          'Sticky' cutter
T   T   C   G   A │ A
```

```
G   G │ C   C
                  'Blunt' cutter
C   C │ G   G
```

Types of cut made by different restriction enzymes

Use of restriction enzymes to insert a gene into a plasmid

Before the introduction of PCR, DNA could be amplified by inserting it into a bacterial plasmid. When the bacterium containing the plasmid divides, the plasmid (and its inserted DNA) is copied. Today, inserting genes into bacterial plasmids is used more to express the gene concerned and to collect the product made, e.g. human insulin. Genes of interest are usually identified by the use of DNA probes, and cut out from the sample of DNA using restriction enzymes. Many eukaryotic genes contain introns (non-coding regions), so this method would also remove the introns and the gene would not be expressed.

1 2nd marker e.g. Lac Z, which is disrupted when plasmid is cut open

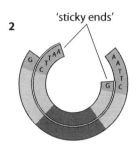

plasmid vector

ampicillin resistance gene

2 'sticky ends'

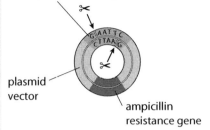

3

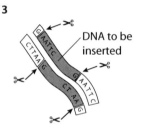

DNA to be inserted

4

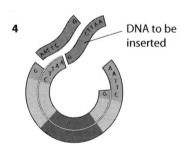

DNA to be inserted

5

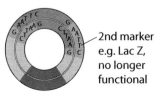

2nd marker e.g. Lac Z, no longer functional

Inserting a gene into a bacterial plasmid

1. The bacterial plasmid contains two marker genes: the first one is for ampicillin resistance so any bacteria that contain the plasmid can grow on an agar plate with ampicillin on it, and their growth is used to confirm that bacteria have taken up the plasmid. The second marker uses a gene which is rendered non-functional if DNA is successfully inserted into it, and is used to confirm insertion of target gene.

2. The plasmid is cut with a restriction enzyme to open the plasmid.

3. The foreign DNA or gene is cut with the same restriction enzyme to ensure complementary sticky ends.

4. DNA is inserted using **DNA ligase** enzyme which joins the sugar-phosphate backbones of the two sections of DNA together.

5. To ensure that bacteria have a plasmid with the donor gene in, the second marker gene is used, e.g. Lac Z gene. The Lac Z gene metabolises x-gal turning it from colourless to blue. Plasmids with an inactive Lac Z gene (and hence containing inserted DNA) will appear blue if x-gal is spread on the plate as they are unable to metabolise it.

Key Term

DNA ligase: a bacterial enzyme that joins sugar-phosphate backbones of two molecules of DNA together.

Grade boost

It is important to know the function of the two marker genes.

quickfire

㉞ Explain the function of the second marker gene in the plasmid.

4.5

The diagram below shows a bacterial plasmid 4400 bases long, and the positions where six different restriction enzymes cut it, e.g. *Pst1* site is found at 35 base pairs from the *EcoR*1 site. Using the diagram answer the following questions:

a) If the plasmid was cut with all six restriction enzymes:

   i How many fragments would be produced?

   ii What is the size of the smallest fragment?

   iii What is the size of the fragment produced by EcoR1 and Bal1?

b) One gene was found between 1515 and 2140 bases in the plasmid. Which restriction enzymes would you have to use to remove the gene with as few extra bases as possible?

>> **Pointer**

If the enzyme cuts a straight section of DNA once, it produces two fragments. If you cut a circular piece of DNA once, you get one linear fragment.

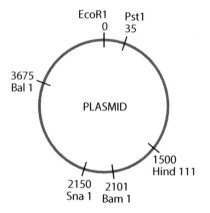

Bacterial plasmid showing restriction enzyme cutting sites

c) If the following DNA sequence was cut with Bam 1 and Pst 1, how many fragments would be produced? The recognition sequences for Bam 1 and Pst 1 are shown. Explain your answer showing the fragments produced.

GATTCCCTAGGATCGAAGTCGGGTTTAAA
CGAAGGGATCCTAGCTTCAGCCCAAATTT

Bam 1	C|CTAGG	Pst 1	G|ACGTC
	GGATC|C		CTGCA|G

Using reverse transcriptase

Many eukaryotic genes contain introns, and are difficult to locate within 3.6 billion bases spread across 46 chromosomes. To solve this problem, scientists turned to another enzyme, **reverse transcriptase** which produces complementary or copy DNA (cDNA) from a mRNA template. By targeting β cells in the pancreas, there is a high proportion of mature mRNA that codes for insulin, which can be extracted and reverse transcribed to cDNA. To ensure proper expression in bacteria, the human regulator sequence (which controls gene expression) is replaced by a bacterial regulator and the cDNA inserted into the plasmid using restriction enzymes and ligase as shown below. Once expressed in bacterial cells, the insulin can be purified for use.

> ### Key Term
> **Reverse transcriptase**: an enzyme that produces DNA from a RNA template.

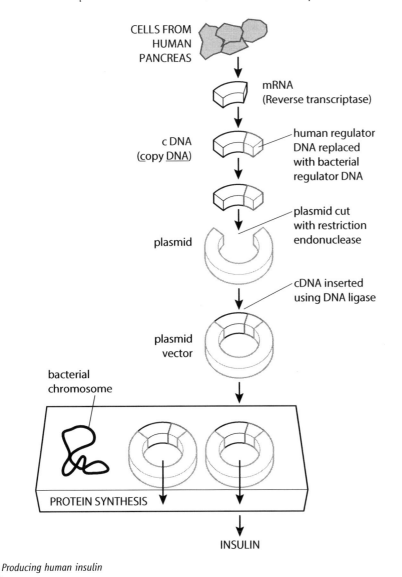

Producing human insulin

- CELLS FROM HUMAN PANCREAS
- mRNA (Reverse transcriptase)
- cDNA (copy DNA)
- human regulator DNA replaced with bacterial regulator DNA
- plasmid cut with restriction endonuclease
- plasmid
- cDNA inserted using DNA ligase
- plasmid vector
- bacterial chromosome
- PROTEIN SYNTHESIS
- INSULIN

> ## ≫ *Pointer*
> Prokaryotes don't have introns in their DNA so this procedure gives intron-free DNA for insertion.

> ### quickᖴire
> ㉟ What is cDNA and how is it made?

> ### quickᖴire
> ㊱ What is the function of DNA ligase in the production of human insulin?

> ### quickᖴire
> ㊲ Suggest one advantage of using mRNA rather than DNA when producing human insulin.

Grade boost

Be careful with pros, cons and hazards: A hazard is more serious than a disadvantage!

Pointer

Remember one pro, one con and one hazard for GM.

Pros and cons of genetically engineering bacteria

Advantages:

- Allows production of complex proteins or peptides which cannot be made by other methods.
- Production of medicinal products, e.g. human insulin, factor VIII clotting factor. These are far safer than using hormones extracted from other animals or from donors. Many people with haemophilia within the UK were infected with HIV during the 1980s from contaminated factor VIII extracts.
- Can be used to enhance crop growth – GM crops.
- GM bacteria have been used to treat tooth decay as they outcompete the bacteria which produce lactic acid that leads to dental caries.

Disadvantages:

- It is technically complicated and therefore very expensive on an industrial scale.
- There are difficulties involved in identifying the genes of value in a huge genome.
- Synthesis of required protein may involve several genes each coding for a polypeptide.
- Treatment of human DNA with restriction enzyme produces millions of fragments which are of no use.
- Not all eukaryote genes will express themselves in prokaryote cells.

Hazards:

- Bacteria readily exchange genetic material, e.g. when antibiotic resistance genes are used in *E. coli* these genes could be accidentally transferred to *E. coli* found in the human gut, or other pathogenic bacteria.
- The possibility of transfering oncogenes by using human DNA fragments thus increasing cancer risks.

GM crops

The use of GM crops is widespread in the USA and is expanding in the EU, Brazil, India and other countries. The advantages and disadvantages of each GM crop are not fully proven. The most widely grown genetically modified crop is soya. Examples of GM crops include:

- Insect-resistant crops using a gene which codes for a toxin from the bacteria *Bacillus thuringiensis (Bt)*.
- Crops tolerant to herbicides like glyphosate (Roundup™) or glufosinate ammonium (Liberty™).
- Crops with stacked traits of both Bt insect resistance and herbicide tolerance.
- Virus resistant crops.

Transforming plants with *Agrobacterium tumifaciens*

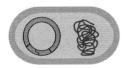

1. Plasmid extracted from the *A. tumifaciens*.

2. Restriction enzyme is used to cut the plasmid and remove the tumour-forming gene.

3. A section of DNA containing a gene for disease resistance is located and isolated using the same restriction endonuclease.

4. The gene is inserted into the plasmid, replacing the tumour-forming gene. DNA ligase is used to join the donor and vector DNA together.

gene for disease resistance ⟶ The plasmid is inserted back into the bacterium.

5. The bacterial cell is introduced into the plant cell. The bacterial cell divides and gene is inserted into the plant chromosome.

6. Transgenic plant cells are grown in tissue culture and transformed plants are regenerated.

Making disease-resistant transformed plants

Benefits of GM crops:

- Increased growth rates, e.g. high crop yields have been reported for cotton and soya, but may not have been sustained.
- Improved nutritional value, e.g. high vitamin A precursor content in Golden Rice.
- Increased pest resistance, e.g. Bt-maize.
- Ease of management, e.g. use of weed killers on resistant crops.
- Tolerance of unfavourable conditions, e.g. drought-resistant cotton and corn crops are being developed.

Concerns with GM crops:

- Genetic contamination, e.g. pollen from GM crops being transferred to other crops which can lead to the development of superweeds e.g. glufosinate – resistant oilseed range crossbred with the weed Charlock to produce a herbicide-resistant weed.
- Misuse of pesticides, e.g. overuse of the weed killer roundup on Roundup™-ready soya.
- Control of agriculture, by corporations, e.g. the control of seed supplies to farmers.

Hazards:

- Threats to biodiversity from the transfer of GM pollen to wild plants which can change natural gene pools. This may result in a reduction in biodiversity.
- Unknown effects of eating new protein produced in crops.

Grade boost

Expect to be asked to explain the risks associated with producing GM crops including the impact upon the environment.

quickfire

(38) What is the main concern regarding the pollen from GM crops?

Gene screening and gene therapy

Genetic diseases involve single gene conditions, e.g. Duchenne muscular dystrophy and cystic fibrosis, chromosomal disorders, e.g. Down's, and multifactorial conditions where faulty genes are part of the cause, e.g. Alzheimer's disease.

Gene screening

Screening for genetic conditions allows for accurate diagnosis and treatment, identification of people at risk of preventable conditions, pre-symptomatic testing for adult-onset disorders, e.g. Alzheimer's disease, and can help families to plan to avoid passing on conditions to children, e.g. Tay-Sachs disease, which is prevalent in Ashkenazi Jews. Screening can involve testing parents, IVF embryos prior to implantation, foetuses during pregnancy (pre-natal testing) and newborns. Screening involves sessions with a genetic counsellor so that the implications are fully understood. Concerns exist over the storage and use of genetic test information, e.g. provision of health care or life insurance, and that people may be discriminated against.

Genetic therapy

The aim of gene therapy is to treat a genetic disease by replacing defective alleles in a patient with copies of a new DNA sequence, but treatment can also involve replicating the function of genes using drugs.

Two possible methods:

 Pointer

Only limited success has been made with somatic cell therapy.

1. Somatic cell therapy – the therapeutic genes are transferred into the somatic (body) cells, of a patient. Any modifications and effects will be restricted to the individual patient only, and will not be passed on through gametes. DNA is introduced into target cells by a vector, e.g. plasmid or virus. For example, in the use of liposomes containing copies of the normal allele, to treat cystic fibrosis.

 Grade boost

Only germ line therapy is permanent, but it is highly controversial.

2. Germ line therapy – sperm or eggs are modified by the introduction of functional genes, which are integrated into their genomes. This would allow the therapy to be heritable and passed on to later generations. This is rare due to ethical and technical reasons.

Duchenne muscular dystrophy (DMD) is a recessive, sex-linked form of muscular dystrophy affecting up to one in 3500 live male births. It is caused by a mutation in the dystrophin gene resulting in the failure to produce dystrophin, an important structural component of muscle tissue. The result is severe wasting of the muscles and sufferers are often wheelchair bound by the time they reach teenage years, and life expectancy is only 27. A drug called drisapersen has been developed which aims to treat DMD by introducing a 'molecular patch' over the exon with the mutation making the gene readable again. A shorter form of dystrophin is produced, but one thought to be more functional than the untreated version. This type of treatment is called exon skipping.

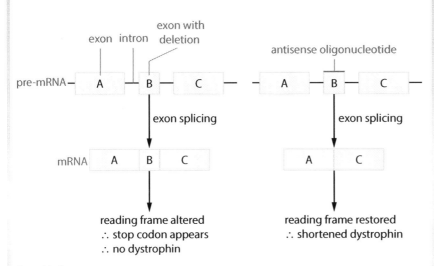

Exon skipping

Cystic fibrosis is caused by a recessive allele that codes for a mutated form of the cystic fibrosis trans-membrane regulator (CFTR). The result is that the membrane protein is unable to transport chloride ions out of cells, and mucus found covering the epithelial tissues remains thick and sticky because the water potential is not lowered by the presence of chloride ions which draw water into the mucus by osmosis. Patients with cystic fibrosis therefore suffer from thickened mucus which blocks bronchioles and alveoli leading to recurrent infections. Mucus also blocks pancreatic ducts leading to poor digestion of food. Gene therapy has involved insertion of the healthy CFTR gene into a **liposome** which is then administered by means of an aerosol. The liposomes fuse with cell membranes lining the bronchioles allowing DNA to enter the cell and be transcribed. As epithelial cells are continuously replaced, this is only a treatment and has to be repeated.

> ## Key Term
>
> **Liposome**: a hollow phospholipid sphere used as a vesicle to carry molecules into a cell.

> ## quickfire
>
> ㊴ Explain why a modified common cold virus could also be used to introduce copies of the normal CFTR allele.

Effectiveness of gene therapy

Results are variable, and therapy often has to be repeated, so it is not a cure. The limited success is due in part to the plasmid often not being taken up, and when it is, the gene it contains is not always expressed.

Genomics and healthcare

Genomics is the study of the structure, function, evolution and mapping of genomes, e.g. the Human Genome and 100K Projects. Genomics should enable healthcare to be improved by:

- More accurate diagnosis of disease.
- Better prediction of the effect of drugs and improved design of drugs. Individual patients metabolise drugs in different ways, so it is important to know whether a drug will be effective, and if so what the safe effective dose would be.

- New and improved treatments for disease as a result of better understanding the biochemistry of diseases, i.e. the faulty proteins produced.

With the introduction of NGS technology it may be possible to look at tailoring therapies to individual patients where an individual could have a unique treatment for a common disease.

Tissue engineering

The first licensed engineered tissue in 1998 was an artificial skin called 'Apligraf' used in place of skin grafts for burns patients. Fibroblasts were removed from skin cells and their life extended by elongating the telomeres present that usually shorten with successive cell division and therefore limit cell mortality. These cells were 'seeded' onto a scaffold, an artificial structure which can support growth of a 3D tissue. Scaffolds must allow diffusion of nutrients and waste products, allow cells to attach and move, and be able to be degraded and absorbed by the surrounding tissues as it grows.

Tissue culture is used to grow large numbers of genetically identical cells quickly from a single parent cell, and is referred to as therapeutic cloning. The main advantage of this is that if patient's cells are used, rejection of the tissue is unlikely. During the tissue culture process, adequate oxygen and nutrients must be provided, and optimal conditions, e.g. temperature and humidity must be maintained. Reproductive cloning of humans involving whole organisms is prevented by law in the UK.

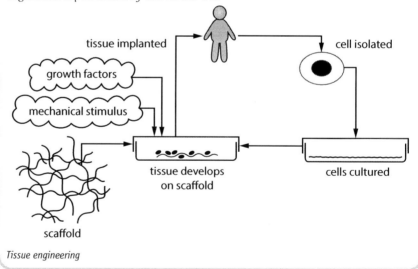

Tissue engineering

quickfire

⑩ What is the function of the scaffold in tissue engineering?

Stem cells

Stem cells are undifferentiated cells which can develop into numerous different cell types given the correct trigger.

The main sources of stem cells are:

- From embryos (embryonic stem cells).
- Adult stem cells, e.g. bone marrow which give rise to new blood cells, but they are not 'true' stem cells as they are pluripotent and cannot differentiate into all types of cell like totipotent stem cells.

Stem cells can be used to regenerate tissues and organs, e.g. pancreatic cells that fail to release sufficient insulin in patients with diabetes, damaged spinal cord cells or skin replacement for burns victims. They can also be used to screen new drugs, and develop model systems to study growth and birth defects.

Advantages of using stem cells:

- Can be produced quickly.
- Produced on a large scale.
- Production of genetically identical cells for transplant, reducing risk of rejection.

Disadvantages of using stem cells:

- In mammals the technique is very expensive and unreliable.
- In plants, disease or entry of pathogens may cause problems.
- Inadvertent selection of disadvantageous alleles.
- Long-term/unforeseen effects such as premature aging.

There are ethical issues associated with obtaining stem cells from embryos and the cloning of human tissues and organs.

Key Term

Stem cell: an undifferentiated cell capable of dividing to give rise to daughter cells which can develop into different types of specialised cell or remain as undifferentiated stem cells.

≫ Pointer

Ethics are a set of standards that are followed by a particular group of individuals, designed to regulate behaviour, i.e. they determine what is acceptable.

Grade boost

The use of cloning and stem cells is controversial. Be prepared to discuss the arguments for and against their use.

quickfire

④ State the difference between adult and embryonic stem cells.

A. Immunology and disease

A disease is an illness of people, animals or plants which is caused by infection, e.g. by a bacterium, fungus or virus, or failure of health, e.g. degeneration of nervous tissue. Disease can affect even the smallest microorganisms, e.g. bacteriophages infect bacteria, e.g. *E.coli*.

Grade boost

Learn your Key Terms in this topic, and how the different types of diseases listed are transmitted and treated.

Key Terms

Endemic: a disease occurring frequently, at a predictable rate, in a specific location or population.

Toxin: a small molecule, e.g. a peptide made in cells or organisms, that causes disease following contact or absorption. Toxins often affect macromolecules, e.g. enzymes, cell surface receptors.

Carrier: an infected person, or other organism, showing no symptoms but able to infect others.

Disease reservoir: the long-term host of a pathogen, with few or no symptoms, always a potential source of disease outbreak.

Infection: a transmissible disease often acquired by inhalation, ingestion or physical contact.

Antigenic type: different individuals of the same pathogenic species with different surface proteins, generating different antibodies.

Epidemic: the rapid spread of infectious disease to a large number of people within a short period of time.

Pandemic: an epidemic over a very wide area, crossing international boundaries, affecting a very large number of people.

Antigen: a molecule that causes the immune system to produce antibodies against it. Antigens include individual molecules and those on viruses, bacteria, spores or pollen grains. They may be formed inside the body, e.g. bacterial toxins.

Vector: a person, animal or microbe that carries and transits an infectious pathogen into another living organism.

Diseases

Disease	Symptoms	Transmission	Treatment
Cholera is caused by the Gram-negative bacterium *Vibrio cholera*. **Endemic** in some parts of the world.	Whilst infected with cholera, the bacteria release a **toxin** resulting in watery diarrhoea and dehydration.	People become infected when they consume contaminated food or water. They become **carriers** of the disease acting as **reservoirs** of the **disease**.	Dehydration is treated by giving clean water and electrolytes, and the **infection** can be treated with antibiotics. Cholera can be prevented by having better sewage and water treatment, safe handling of food, and by washing hands. A vaccine is available.
Tuberculosis (TB) is caused by the bacillus *Mycobacterium tuberculosis*.	Cells in the lungs become damaged, forming tubercles or nodules. Patients present with chest pain, blood in their sputum, and a fever. If left untreated it can be fatal due to extensive lung damage.	Infection spreads rapidly by inhalation of water droplets from coughs and sneezes from infected people. It spreads more rapidly amongst people with depressed immune systems, e.g. from HIV-AIDS, and in crowded conditions.	Treatment is by a long course (6 months) of antibiotics, but some strains are now showing antibiotic resistance. A BCG vaccine (made from a weakened strain of a related bacterium *M. bovis*) is available.
Smallpox is caused by the virus *Variola major*.	It causes fever and pain with a rash and fluid-filled blisters. It can leave patients with blindness and limb deformities.	Infection enters small blood vessels in the skin and mouth, quickly spreading throughout the body.	Pain killers and fluid replacement therapy help relieve the symptoms. Antibiotics can help treat secondary infections but mortality is high at 30%. A vaccine made from the *Vaccinia* virus resulted in the eradication of smallpox by 1979.

Disease	Symptoms	Transmission	Treatment
Influenza virus has three subgroups which contain viruses with different **antigenic types** (they have different antigens on their surface) meaning that the immune system is unable to produce adequate protection from all types of flu, which can result in **epidemics**. **Pandemics** can occur, e.g. Spanish flu in 1918–20 which killed over 50 million people worldwide.	Flu virus attacks the mucous membranes in the upper respiratory tract resulting in fever, sore throat and cough. Secondary infections can result in vulnerable people, e.g. children and the elderly.	Flu spreads by droplets from coughs and sneezes.	Infection spread can be reduced by regular washing of hands, using tissues to catch sneezes and coughs and isolating patients. Vaccines have some effect depending upon the degree of mutation in the viral **antigens**. Antiviral drugs, e.g. Tamiflu, have some success in reducing length of symptoms and may prevent flu from developing if taken as a preventative measure.
Malaria is caused by a protoctistan parasite *Plasmodium*. The two species that cause the largest numbers of cases are *P. falciparum* and *P. vivax*, which are transmitted by over 100 different species of *Anopheles* mosquitoes. It is the female mosquito that acts as the **vector** when it feeds on blood. Malaria is endemic in some sub-tropical regions and can become epidemic during wet seasons.	The cycle of red blood cells bursting repeats every few days and gives rise to recurring fevers.	When a female mosquito feeds on blood which it ingests the *Plasmodium* parasite which migrates to the liver where they develop before being released to infect red blood cells causing them to burst.	Malaria is treated using a variety of anti-malarial drugs, the first of which was quinine. *Plasmodium* now shows resistance to many of these drugs, so drug combinations are commonly used. Prevention by using mosquito repellants and insecticide impregnated bed nets are seen as preferable in the fight against malaria. Other methods of controlling mosquitoes have been used, for example use of insecticides, draining of stagnant water where mosquitoes lay their eggs, releasing sterile male mosquitoes and biological control using fish that eat mosquito larvae.

Virus pathogenicity and reproduction

Viruses are intracellular parasites that use the host cell's metabolic pathways to reproduce. They cause pathogenic effects to the host in several ways:

- During cell lysis, the cell bursts allowing virus particles to emerge and infect other cells, which causes many of the symptoms seen.
- Viruses produce many different toxic substances, e.g. viral proteins can inhibit DNA and protein synthesis, and glycoproteins produced by the *Herpes* virus can cause cell fusion.
- Cell transformation can occur, which means that viral DNA integrates into the host chromosome. If this occurs into a proto-oncogene or tumour suppressor gene, uncontrollable cell division can result (cancer), e.g. human papilloma virus which has been shown to cause cervical cancer.
- Immune suppression results from some viruses, e.g. HIV virus, which destroys T helper cells.

Control of bacterial infection

- Sterilisation results in all microorganisms and spores being killed. This is usually achieved by autoclaving in excess of 121°C for at least 15 minutes or by use of gamma radiation.
- Disinfection with antiseptics or disinfectants will remove the majority of microorganisms but not all.
- **Antibiotics** may be either:
  - Bactericidal and so kill bacteria, e.g. penicillin
  - Bacteriostatic and so prevent the growth of bacteria within the body, e.g. tetracycline which inhibits protein synthesis. Tetracycline only prevents growth whilst the antibiotic is present.

> **Key Term**
>
> **Antibiotic**: a medicine which inhibits growth or destroys bacteria.

quickfire

㊷ State the difference between sterilisation and disinfection.

Bacterial cell wall

The structure of the bacterial cell wall greatly influences the type of antibiotic that will be effective. Gram-positive bacteria have thicker cell walls containing peptidoglycan with polysaccharide molecules cross-linked to amino acid side chains. The cross-linking confers strength and protects against osmotic lysis. Gram-negative bacteria have thinner but more complex cell walls. They still have a little peptidoglycan, but this is covered by a layer of lipoprotein and lipopolysaccharide, which protects the bacteria from some antibacterial agents such as lysozyme and penicillin.

Penicillin prevents the synthesis of the cross-links in peptidoglycan, because transpeptidase enzymes that cross-link the polysaccharide molecules to the amino acid side chains are inhibited by penicillin. This weakens the cell wall, and as water enters by osmosis, the cell bursts (osmotic lysis). Because penicillin only causes lysis in Gram-positive bacteria penicillins are referred to as narrow spectrum antibiotics.

Tetracycline and chloramphenicol work differently as they stop protein synthesis within the bacterial cell, but do not affect normal cell metabolism. Tetracycline works by binding to the 30S subunit of the bacterial ribosome in the second position blocking further tRNA attachment. As tetracycline binds reversibly, the effect is bacteriostatic. Because tetracycline affects both Gram-positive and negative bacteria it is referred to as a broad spectrum antibiotic.

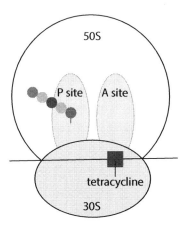

Tetracycline mechanism

Antibiotic resistance

Antibiotic resistance is the ability of a microorganism to withstand the effects of an antibiotic. It evolves naturally as bacteria have a high mutation rate and divide rapidly. If a mutation arises which changes one allele within the bacterium's plasmid, resulting in the bacterium being able to produce an enzyme which can break down the antibiotic, the bacterium has a selective advantage in the presence of antibiotics, so passes the advantageous allele on by plasmid transfer or in asexual reproduction. MRSA (Methicillin Resistant *Staphylococcus aureus*) is a bacterium that is resistant to beta-lactam antibiotics, including methicillin, oxacillin, penicillin, and amoxicillin. Resistance has arisen largely through the over use of antibiotics for viral infections and prevention of infection in farmed animals. Antibiotics are ineffective against viruses because they do not have the metabolic pathways present or possess any peptidoglycan, but any bacteria present would be exposed to the antibiotic and so could start to develop resistance.

Immune response

Primary defences against entry of pathogens in animals include:

- Skin – a physical barrier and has a slightly acidic pH preventing growth of pathogens.
- Natural skin flora – offer protection by competing with pathogenic bacteria and unlike these bacteria, the flora is not easily removed by washing.
- Lysozymes – produced in tears which can hydrolyse bacterial cell walls.
- Stomach acid – kills bacteria.
- Cilia and mucus – in the trachea and other mucous membranes these trap and remove particles and microbes from the air.
- Blood clotting – seals open wounds.
- Inflammatory response – increases blood flow to the site of injury.

Lymphocytes are a type of cell found in the blood that originate from stem cells in the bone marrow of the long bones. Every individual possesses many types of lymphocytes and each lymphocyte is capable of recognising one specific antigen.

There are two types of lymphocytes involved in the immune response:

- B-lymphocytes which mature in the spleen and lymph nodes.
- T-lymphocytes which are activated in the thymus gland. There are three types of T-lymphocytes:
  - Killer (cytotoxic) T cells which bind to foreign cells with complementary antigens and destroy them.
  - Helper T cells that stimulate phagocytosis and antibody production and activate T killer cells.
  - T memory cells which remain in the blood and can respond quickly in case the same infection is encountered again.

The immune response is the body's reaction to a substance which is recognised as non-self, e.g. a foreign antigen. There are two components to the immune response:

1. The humoral response which results in the production of **antibodies** by the B-lymphocytes. When a B-lymphocyte recognises its specific antigen it divides rapidly to produce clones which then change into two cell types:
   i) Plasma cells which are short lived and immediately secrete antibodies.
   ii) Memory cells which live for considerably longer and give rise to the secondary immune response if the same infection is encountered again.

The variable region on the antibody is specific to each antigen and acts as the antigen binding site, allowing each antibody to bind to two antigen molecules Microbes with antigens on their surfaces clump together (agglutinate) making it more difficult for them to infect other cells and more easy for macrophages to engulf them.

 Pointer

A good way to remember where the T-lymphocytes are activated is to remember T = thymus.

Key Term

Antibody: an immunoglobulin produced by the body's immune system in response to an antigen.

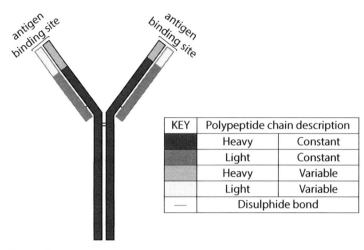

An antibody molecule

>> **Pointer**

Antibodies are proteins.

KEY	Polypeptide chain description	
	Heavy	Constant
	Light	Constant
	Heavy	Variable
	Light	Variable
—	Disulphide bond	

2 The cell-mediated response involves attacking foreign material inside cells, e.g. a virus, and phagocytes, B-lymphocytes and T-lymphocytes are all activated. T-lymphocytes respond to specific antigens on the surface of cells and divide rapidly by mitosis to form clones. Three types of cells are involved:

i) T helper cells which cooperate with B-lymphocytes to initiate antibody response. They release chemicals including cytokines which stimulate phagocytes to engulf pathogens and digest them.

ii) T killer cells which engulf and lyse target cells.

iii) T memory cells which remain in the blood in case of reinfection.

Types of immune response

1. Primary response

During the latent period, the body reacts to a foreign antigen by producing antibodies. The process involves:

- antigen presenting cells (including macrophages) carry out phagocytosis and incorporate foreign antigen into their cell membranes.
- T helper cells detect these antigens and secrete cytokines, which stimulate B cells and macrophages.
- B cells are activated and undergo clonal expansion to produce plasma cells and memory cells.
- Plasma cells secrete antibodies.
- Memory cells remain in the blood to protect against reinfection.

>> **Pointer**

Memory cells confer long-term protection.

2. Secondary response

If the body receives a second exposure to the same antigen, memory cells are stimulated to clone themselves and produce plasma cells, which produce antibodies. This response is much more rapid than the primary response and produces up to 100 times the concentration of antibodies, which remain in the blood for longer than with the primary response.

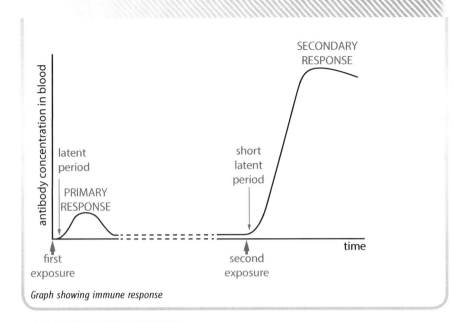

Graph showing immune response

Types of immunity

1. Passive immunity

   This occurs when the body receives antibodies, either naturally (e.g. from mother's milk or via the placenta) or artificially from an injection where rapid protection is needed, e.g. when exposed to *Rabies* virus, antibodies can be given. The advantage is that the body receives immediate protection but the disadvantages are that the protection is short-lived because the body has not produced memory cells, and the injection of artificial antibodies may be perceived by the body as foreign (non-self) and so it may make antibodies against them.

2. Active immunity

   This occurs when the body produces its own antibodies in response to antigens being present. This protects against reinfection where the antigens on the invading microorganism are the same. Antigen-specific memory cells are produced and some antibodies remain in the blood to protect against reinfection. Active immunity can be natural when the person suffers a disease and makes antibodies against it or artificial when antigens are supplied artificially in the form of a **vaccine**, which triggers antibody production without the symptoms of the disease.

Immunisation programmes

Vaccination programmes are designed to protect populations from harmful diseases. Any antigens used in vaccines must be highly immunogenic to stimulate a protective immune response. Programmes need to be cost effective and take into consideration any possible side effects. Some vaccination programmes are not 100% effective, e.g. influenza, due to antigenic variability within the virus.

Key Term

Vaccines: antigens isolated directly from the pathogen, weakened (attenuated) strains of the pathogen, e.g. MMR, inactive pathogens, e.g. whooping cough or inactivated toxin, e.g. tetanus.

quickfire

⁴³ Why must antigens in vaccines be highly immunogenic?

quickfire

⁴⁴ Explain two differences between active and passive immunity.

B. Human musculoskeletal anatomy

The three main tissues involved in the musculoskeletal system are cartilage, bone and skeletal muscle.

Cartilage

Cartilage is a hard and flexible **connective tissue** that permits movement of structures, e.g. the ribcage, whilst its rigidity provides support to structures, e.g. the trachea. Cartilage consists of chondrocyte cells embedded in a matrix which is secreted by the cells themselves, and are found in spaces called lacunae. The absence of nerves and blood vessels means that if damaged, cartilage takes a long time to heal because nutrients have to diffuse into the matrix. The type of cartilage depends on the presence of collagen fibres which determines it's function.

Cartilage	Structure	Function
Hyaline cartilage	The weakest form of cartilage.	It is found on the articular surface of bones, in the nose and trachea.
Yellow elastic cartilage	Chondrocytes are surrounded by dense elastic fibres and collagen making it elastic but able to retain its shape.	It is found in structures such as the external ear (pinna).
White fibrous cartilage	Sometimes called fibrocartilage, is the strongest cartilage as collagen is arranged into dense fibres increasing the tensile strength.	It is found in intervertebral discs.

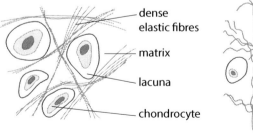

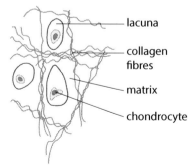

Yellow elastic cartilage

Diagram to show structure of fibrocartilage

Bone

Bone provides structural support, allows for movement due to muscle attachment sites, and is also involved in mineral regulation, e.g. storage of calcium and phosphorus. There are two types of bone:

1. Spongy (cancellous) bone located at the end of the long bones containing bone marrow where blood cells are made.

2. Compact bone represents the majority of bone in the body and consists of cells called osteoblasts which continually build up bone by laying down the inorganic component of the matrix, and cells called osteoclasts which break it down. Cells are contained within a matrix which is:

   - 30% organic consisting mainly of collagen fibres. It is hard and resists fracture.
   - 70% inorganic consisting mainly of hydroxy-apatite containing calcium and phosphate. The inorganic component is hard and resists compression.

Compact bone contains units called **Haversian systems**, which run longitudinally through bone and are supplied with blood by vessels running through Volkmann canals. Concentric rings called lamellae surround the Haversian canals.

Key Term

Haversian system: the structural and functional unit of compact bone; the osteon.

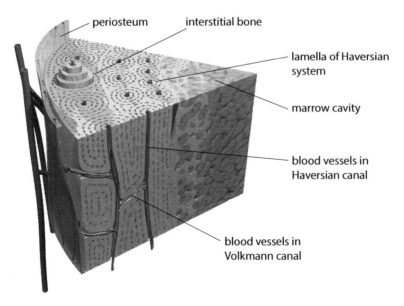

Bone structure

Bone in the vertebra, ribs and limbs is formed from hyaline cartilage in the embryo which ossifies: osteoblasts secrete bone matrix around the cartilage. The cartilage is broken down by osteoclasts, as blood vessels invade. Ossification begins at the ends of the long bones in the limbs. The inorganic matrix is laid down in the direction of the stress placed on the bone which maximises strength.

quickfire

45 State the difference between the roles of osteoblasts and osteoclasts.

Bone diseases

Rickets results from the inadequate deposition of minerals in the growing bones of children as a result of a lack of the fat soluble vitamin D or calcium in the diet. Vitamin D is required for the absorption of calcium in the intestines, and is found in butter, eggs and fish liver oils, and is synthesised in the skin from a precursor (inactive vitamin D) using UV light present in sunlight. Historically the incidence of rickets was due to poor diet, but more recently it has increased due to the use of sun blocks, and spending too much time indoors. In adults where bones have finished growing, a milder form of the disease exists called **osteomalacia**.

Brittle bone disease called *osteogenesis imperfecta*, occurs in about 1 in 20,000 live births. It is a genetic condition caused by a mutation in the gene responsible for making collagen, which results in collagen not coiling as tightly as in healthy individuals, making bones more susceptible to fracture. It can be treated by the use of drugs to increase mineral density in bones, physiotherapy, or surgery to place metal rods in the long bones.

Osteoporosis is the abnormal loss of density in spongy and compact bone which results in increased risk of fractures. Incidence increases with age due to decreasing levels of oestrogen and testosterone, but is more common in people with a family history, or those who smoke and drink alcohol. Bone density can be improved by increasing levels of exercise, using drugs to enhance calcium uptake, and diets rich in calcium and vitamin D.

Skeletal muscle

Skeletal muscle is made up of muscle fibres which are long thin cells containing many nuclei. Each fibre contains many **myofibrils**. Under the microscope skeletal muscle appears with transverse stripes called striations.

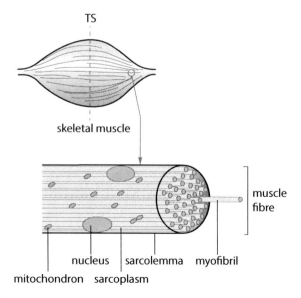

A muscle fibre

Myofibrils are packed together surrounded by the sarcoplasmic reticulum. Many mitochondria are present, which provide ATP for muscle contraction. Myofibrils are made from the proteins actin, myosin, troponin and tropomyosin which are organised into **myofilaments**. The striped appearance is due to the **ultra-structure** of the myofibrils. Myofibrils are arranged into repeating units called sarcomeres each 3μm long: the dark band (A band) forms from the alignment of thick myosin filaments and a light band (I band) forms from the alignment of thin actin filaments in adjacent myofibrils. Z lines mark the end of each sarcomere and allow attachment of the actin filaments.

The T system refers to T tubules (transverse tubules) which cross the myofibrils. They are formed from infoldings of the sarcolemma and transmit nervous impulses through the muscle fibre very quickly enabling all the myofibrils to contract at the same time.

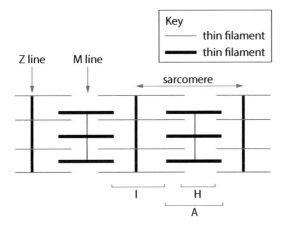

Portion of a myofibril showing its ultra-structure

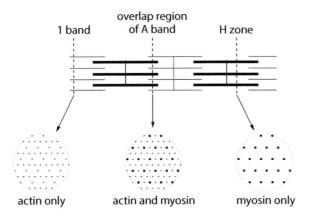

Section through a myofibril showing arrangement of myofilaments

Key Terms

Myofilament: thin filaments of mainly actin and thick filaments of mainly myosin, in myofibrils, that interact to produce muscle contraction.

Ultra-structure: the detailed structure of a cell as seen in the electron microscope. It is also called fine structure.

Grade boost

You can remember acTin as 'thin' strand.

Key Term

Sliding filament theory: the theory of muscle contraction in which thin, actin filaments slide between thick myosin filaments, in response to a nervous impulse mediated by the T system.

quickFire

46 Which sections of a sarcomere shorten during muscle contraction?

Sliding filament theory of muscle contraction

The **sliding filament theory** describes how the actin filaments slide between the myosin filaments to shorten the length of the sarcomere.

Actin contains three different proteins:

- G-actin are globular proteins that are joined into a long chain. Two of these chains twist around each other forming F-actin, which is a fibrous strand.
- Tropomyosin wraps around F-actin in a groove between the two chains.
- Troponin is a globular protein located at intervals along the F-actin molecule.

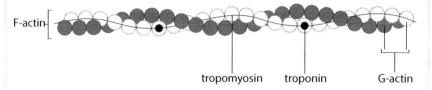

The actin complex

Whereas myosin molecules have:

- A globular head and a fibrous tail.
- Each head is 6nm apart from the head on the adjacent myosin molecule.

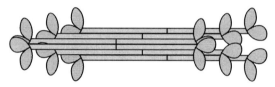

Myosin molecules in a thick filament

During contraction:

- Sarcomeres shorten and so myofibrils and muscle shorten.
- I band shortens.
- H zone shortens.
- A band remains the same length.

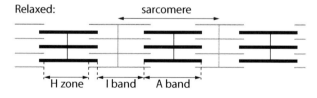

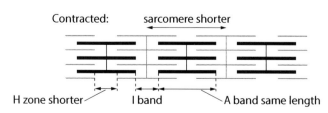

Sliding filament theory

Contraction of muscles fibres is described by the sliding filament theory and involves the following:

1. Once a nervous impulse arrives at the neuro-muscular junction, the wave of depolarisation passes across the muscle fibre via the T tubules.
2. Ca^{2+} ions are released from the sarcoplasmic reticulum.
3. Ca^{2+} ions bind to troponin, changing the shape of the troponin molecule.
4. This causes tropomyosin to change position exposing the myosin binding sites on the actin.
5. Myosin heads form cross-bridges with the myosin binding sites on the actin.
6. The myosin head bends pulling the actin past the myosin. This is referred to as the power stroke.
7. ATP at the end of the myosin head is hydrolysed into ADP and Pi which are released.
8. The cross-bridge is broken when ATP attaches to the myosin head which returns to its original position.
9. More ATP is hydrolysed to ADP and Pi and a cross-bridge forms with the thin filament further along.
10. The process continues until the Ca^{2+} ions are actively pumped back into the sarcoplasmic reticulum.

>> **Pointer**
Remember that actin filaments slide between the myosin.

>> **Pointer**
ATP is needed to break the cross-bridges.

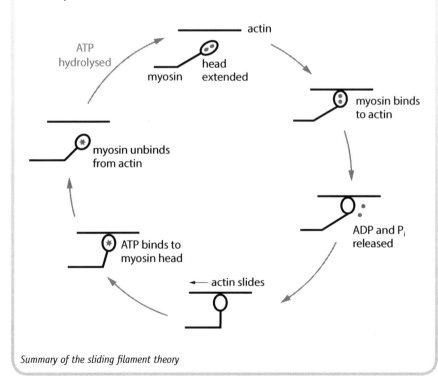

Summary of the sliding filament theory

Types of muscle fibres

There are two types of muscle fibre:

1. Slow twitch, which contract more slowly than fast twitch fibres but with less force.
   - They have *more* mitochondria than fast twitch fibres so are adapted for *aerobic* respiration.
   - Adapted for continuous extended contraction over a long period of time.
   - Have a high resistance to fatigue.
   - Have a rich blood supply, high numbers of mitochondria, high myoglobin levels but low resistance to lactic acid.
   - Have a lower density of myofibrils.
   - Higher proportions found in, e.g., marathon runners.

2. Fast twitch, which contract quickly with more force.
   - Have fewer mitochondria so are adapted for *anaerobic* respiration.
   - Generate short bursts of strength / speed.
   - Higher proportions found in, e.g., sprinters and body builders.

The proportions of slow and fast twitch fibres can be altered by training.

quickfire

㊼ How are slow twitch fibres adapted for aerobic respiration?

4.B

Complete the following table to show the effects of training. The first one has been done for you.

Type of training	Effect of training	Advantage
Endurance training	Increase in number and size of mitochondria	More aerobic respiration possible
Endurance training	Capillary network increases	
Weight training	Increase in number of myofibrils and size of muscles	
Endurance training	Increase in amount of myoglobin	
Weight training	Increase in tolerance to lactic acid	

Exercise

During exercise the main energy source is muscle glycogen which is stored in muscles. Protein is also used as an energy store in preference to fat. Athletes eat high carbohydrate meals the night before a race to increase glycogen stores in muscles. This is referred to as carbohydrate loading.

Under anaerobic conditions the body relies initially upon creatine phosphate stores that were made during aerobic conditions. Creatine phosphate releases its phosphate as oxygen levels fall allowing ADP to be phosphorylated allowing for intense bursts of activity. Once ATP is available again creatine phosphate stores can be replenished. Once both of these have been exhausted, the muscles rely upon anaerobic respiration:

reduced NAD → NAD

pyruvate → lactate

Anaerobic respiration results in the build-up of lactate (lactic acid) in the muscles which causes fatigue and cramp. During cramp, the build-up of lactate inhibits chloride ion effect which results in sustained contraction. An oxygen debt is created as oxygen is needed to break down the lactate.

» Pointer

Remember to revisit your notes on respiration.

Grade boost

Reduced NAD is used to reduce pyruvate to lactate regenerating NAD to allow glycolysis to continue.

Structure and function of the human skeleton

The human skeleton is made up of the axial and appendicular skeleton.

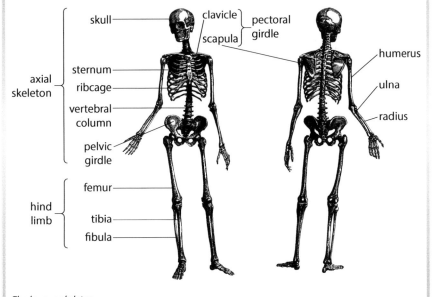

The human skeleton

Fractures

Fractures can be caused by high impact or stresses where the force on the bone exceeds its strength, e.g. through trauma. Fractures are more likely when bone strength is reduced, e.g. through osteoporosis, bone cancer or osteogenesis imperfecta.

Types of fracture:

- Simple or closed fractures do not involve other tissues and are described according to the positions of the bones following the fracture, e.g. comminuted where the bone fragments into several pieces. When the bones are still in their normal position, the fracture is referred to as non-displaced, but when the bones are misaligned the fracture is referred to as displaced.
- Other tissues may become involved, e.g. in open or compound fractures when the bone penetrates the skin or other internal tissues or organs, and this greatly increases the chance of infection.

Treatment:

- Pain is managed by the use of pain-killing medication.
- Broken bones are re-aligned manually if necessary and the area immobilised by the use of a splint or cast. Bones heal naturally as osteoblasts produce new bony tissue to support the broken bone and the osteoclasts remodel the bone.
- Where fractures involve many bone fragments, screws or metal plates can be surgically inserted to support the bones, which can speed up recovery. This is particularly important in hip fractures where continued immobility can lead to further complications, e.g. deep vein thrombosis or pulmonary embolisms.
- Antibiotics may be necessary where surrounding tissue has been damaged.

Vertebral column

The vertebral column consists of 33 vertebrae held in place by muscles, ligaments and tendons:

- 7 cervical vertebrae in the neck which also contain vertebrarterial canals which carry blood vessels.
- 12 thoracic vertebrae in the back to which ribs attach.
- 5 lumbar vertebrae in the lower back.
- 5 sacral vertebrae which have become fused into the sacrum.
- 4 coccyx.

All vertebrae have:

- A vertebral body which is weight bearing. This is largest in the lumbar vertebrae reflecting the additional weight supported.
- Transverse and spinous processes for the attachment of muscles.
- Facet joints which allow articulation with vertebrae above and below, and in the case of thoracic vertebra with the ribs too.
- A vertebral canal which contains and protects the spinal cord. This decreases in width towards the base of the vertebral column which reflects the width of the spinal cord.

Grade boost

Be able to describe the main features of all vertebrae and their functions.

quickfire

48 State one feature present in cervical vertebrae which is absent in other vertebrae.

The shape of the vertebrae and angle of the facet joints and spinous processes vary towards the base of the vertebral column, which allows for varying degrees of movement, for example:

- The cervical spine is capable of wide-ranging rotation, flexion and side bending movements.
- The degree of rotation and extension movements decreases down the thoracic spine.
- Lumbar spine is largely limited to flexion (bending forward) and extension (bending backwards) movements.

Grade boost

The shape of the vertebrae and angle of facet joints determine the range of movement possible.

≫ Pointer

Remember flexion = forward movement.

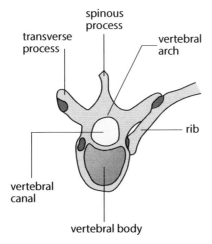

Thoracic vertebrae and rib (anterior view)

Postural deformities

The human spine naturally has a concave curvature to the cervical and lumbar regions described as the S-shape which is easily viewable from the side. In scoliosis there is an additional sideways curve to the spine which can be caused by posture, injury or genetic causes. Scoliosis can be treated by physiotherapy, but in severe cases surgery is needed to brace the spine.

Flat foot occurs when the arch of the foot (instep) fails to develop properly and results in overpronation, which places additional strain on other joints including the spine. It is caused by genetic or congenital factors but can also result from injury as the arch is not fully developed until the age of 10. It can be treated with specialised footwear but in severe cases surgery is an option.

Knock-knees most commonly results from rickets (vitamin D or calcium deficiencies), but can arise from injuries or from a bone infection.

Grade boost

Be able to explain at least three functions of the human skeleton.

Functions of the skeleton

Function	Explanation
Support	Determines the body shape, e.g. thorax is supported by the ribs and sternum.
Muscle attachment	Allows for the attachment of muscles to projections or processes known as origins and insertions.
Protection	Protects many organs, e.g. the ribcage protects the heart and lungs, the cranium protects the brain due to bones being strong and rigid.
Production of blood cells	Bone marrow contains haemopoietic stem cells that produce blood cells. Bone marrow is concentrated in the limbs.
Store of calcium	70% of bone consists of hydroxyl-apatite, and up to 40% of bone is calcium ions.

Joints

A joint is found where bones meet. Joints are classified according to the type of movement possible:

1. Immovable joints or sutures form where bones grow together. The skull consists of 8 cranial bones and 14 facial skeleton bones which begin to fuse together after birth. Some movement in the cranial bones at birth allows the baby's head to fit through the birth canal.

2. Movable joints (synovial joints):

   a. Gliding joints allow bones to glide over each other, e.g. vertebrae, and wrist bones.

   b. Hinge joints allow movement in one plane, e.g. knees, which allows flexion and extension only.

   c. Ball and socket joints, e.g. shoulder joint, which allow movement in more than one plane.

Synovial joints consist of:

- Cartilage on articular surfaces reducing wear.

- Joint capsule consisting of the synovial membrane and ligaments holding joint together.

- A small space called the synovial cavity which contains synovial fluid (secreted by the synovial membrane) acting as a lubricant and shock absorber.

Key Term

Synovial joint: a joint at which bones' movement is lubricated by articular cartilages and synovial fluid, secreted by a synovial membrane. The joint is held in a ligamentous joint capsule.

quickfire

㊾ State the function of the synovial membrane.

quickfire

㊿ State the function of the cartilage in a synovial joint.

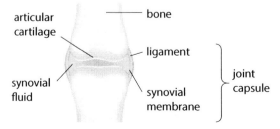

Synovial joint

Antagonistic muscles work in pairs to coordinate movement at joints where a flexor muscle allows for bending and an extensor muscle for straightening, e.g. elbow joint. When one muscle is contracted, the other is relaxed.

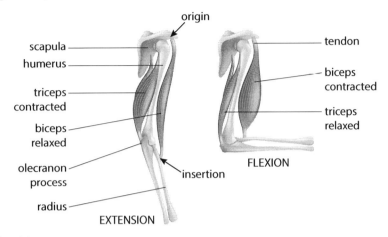

Elbow joint

Tendons consist of 86% collagen fibres arranged in parallel, providing very strong connections between muscles and bone. Due to a lack of elastic fibres, tendons do not stretch, so remain the same length ensuring that all the movement of the muscle is passed on to the bone.

> **Grade boost**
>
> You should be able to label a typical synovial joint, e.g. elbow, and know the functions of each structure.

> **Key Term**
>
> **Antagonistic pair**: when one muscle of an antagonistic pair is contracted the other is relaxed. The contraction and relaxation are coordinated.

Joints as levers

A lever is a rigid, movable structure that pivots about a fixed position known as the fulcrum. In the body, the bone acts as the movable structure, whilst the joint is the fulcrum. The effort is produced by the muscle contracting and the load represents the part of the body moved. There are three types of lever:

> **Grade boost**
>
> Remember FLE. 1st order the fulcrum is in the middle, 2nd order the load is in the middle and 3rd order the effort is in the middle.

Type of lever	How it works	Diagram
1st order lever	Fulcrum is in the middle like a seesaw. In the case of the skull, the effort provided by the neck muscles must balance the load of the skull to maintain posture. This is an example of a force magnifier.	*1st order lever*

| 2nd order lever | Load is in the middle like a wheelbarrow and occurs in certain situations, e.g. standing on tip toes where the toes act as fulcrum and calf muscles provide lift, rather than for maintaining posture. | 2nd order lever |
| 3rd order lever | Effort is in the middle, e.g. the biceps and forearm where the load moves in the same direction as the effort. It is an example of a distance magnifier, i.e. the load moves a larger distance than the effort. | 3rd order lever |

Calculation

When a lever is at equilibrium, $F_1 \times d_1 = F_2 \times d_2$ where F_1 is the force exerted by the load and F_2 is the force exerted by the effort.

Using the second order lever example of standing on tiptoes where:

F_1 is the body weight or load = 100 kg (and assume that 1 kg = 9.8 newtons, N)
d_1 is the distance from the toes (fulcrum) to the heel bearing the weight = 0.20 m
d_2 is the distance from the toes (fulcrum) to the insertion of the calf muscle = 0.23 m

What is the effort needed to stand on tiptoes (F_2)?

$F_1 \times d_1 = F_2 \times d_2$

$(100 \times 9.8) \times 0.20 = F_2 \times 0.23$

$F_2 = \dfrac{(100 \times 9.8) \times 0.20}{0.23}$

$= \dfrac{196}{0.23}$

$= 852.2$ newtons

Joint diseases

Arthritis is a group of conditions in which joints become inflamed. There are two main types:

1. **Osteoarthritis** is the commonest joint disease and is degenerative, and involves the breakdown of articular cartilage faster than it can be replaced as a result of changes in the collagen and glycoprotein. The result is inflammation and joint swelling resulting in pain and joint stiffness. Where bone rubs directly on bone, spurs may form which limit movement. Risk increases with age, and being overweight places more strain on joints, especially the knee. Repeated activity, e.g. flexing of knee during sport also increases risk. No genetic link has yet been identified, and it is not a form of autoimmune disease like rheumatoid arthritis. Treatment involves controlling pain with non-steroidal anti-inflammatory drugs (NSAIDs), and structured exercise. In severe cases joints can be replaced.

   Advantages of joint replacement are:

   - Relief from long term pain.
   - Reduced drug intake.
   - Increased mobility.
   - Restore normal activity and quality of life.

   Disadvantages of joint replacement are:

   - Surgical risks such as increased risk of a blood clot and infection.
   - Long recovery periods.
   - Increased risk of dislocation especially the hip.
   - The replacement joint could fail after 15–20 years and there are increased risks with surgery for a second replacement.

2. **Rheumatoid arthritis** is an autoimmune disorder that attacks bone and cartilage at joints resulting in severe inflammation from increased blood flow, and restricted movement in joints especially in the hands and wrists. Physiotherapy and use of NSAIDs can help reduce inflammation. Risks of developing the disease are higher if someone smokes, has a high consumption of red meat or coffee, but there is also a genetic component.

Key Terms

Osteoarthritis: a degenerative condition in which articular cartilage degrades and produces painful, inflamed joints.

Rheumatoid arthritis: an autoimmune condition in which bone and cartilage at joints is attacked, producing pain, swelling and stiffness.

quickfire

(51) State two differences between osteoarthritis and rheumatoid arthritis.

>> Pointer

More advanced vertebrates have more sophisticated brains with a more developed cerebrum in the forebrain.

quickfire

52 Name the membrane covering the brain surface.

quickfire

53 Match the part of the brain (A–E) with its functions (1–5).
A Cerebellum
B Medulla oblongata
C Cerebrum
D Hypothalamus
E Hippocampus

1 Consolidates memories into a permanent store.
2 Regulates body temperature.
3 Control of heart rate.
4 Controls voluntary behaviour.
5 Maintenance of posture.

C. Neurobiology and behaviour

The brain

The brain is responsible for coordinating responses to sensory stimuli. It is surrounded by three membranes called **meninges**: the pia mater on the brain surface, the arachnoid mater, and the dura mater which is attached to the skull. The brain is continuous with the spinal cord, and contains four **ventricles** (spaces) which are continuous with the central canal of the spinal cord and so contain cerebro-spinal fluid (CSF). CSF supplies neurones in the brain with oxygen and nutrients, e.g. glucose.

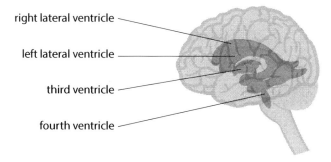

right lateral ventricle

left lateral ventricle

third ventricle

fourth ventricle

Ventricles in the brain

The brain has three regions:

1. Forebrain contains the limbic system, which is associated with memory, learning and emotions consisting of:
   - Hypothalamus, which regulates functions such as sleep, body temperature, thirst, and blood glucose concentration. It releases hormones, e.g. ADH, through the pituitary gland.
   - Hippocampus, which interacts with areas in the brain cortex and is involved in learning, reasoning and personality and also consolidates memories into a permanent store.
   - Thalamus, acts as a relay centre sending and receiving information to and from the cerebral cortex.

   It also contains the **cerebrum** which controls voluntary behaviour and learning, reasoning, personality and memory functions.
2. Midbrain links the forebrain and hindbrain.
3. Hindbrain or reptilian brain which controls basic body functions needed for survival. It includes:
   - **Medulla oblongata**, which is involved with the control of heart rate, ventilation and blood pressure and contains many important centres of the autonomic nervous system.
   - **Cerebellum**, which is involved with the maintenance of posture and the coordination of fine voluntary muscular control needed for writing, and playing a musical instrument, etc.

The autonomic nervous system

The **autonomic nervous system** is the part of the nervous system which controls automatic processes including heart rate, ventilation rate, blood pressure, digestion and temperature regulation. It is divided into:

1. The sympathetic nervous system which generally has excitatory effects on the body, e.g. increasing heart and ventilation rates. Most synapses in the sympathetic nervous system release noradrenaline as the neurotransmitter, which has similar effects on target cells as the hormone adrenaline.

2. The parasympathetic nervous system generally has an inhibitory effect on the body, e.g. decreasing heart and ventilation rates. Most synapses in the parasympathetic nervous system release acetylcholine as the neurotransmitter.

The systems work antagonistically to control body functions.

Control of heart rate

Heart rate is regulated by changes in blood pH which happen during exercise when carbon dioxide builds up in the blood. Control is autonomic by the cardiovascular centre in the medulla oblongata, and requires no conscious thought.

- Heart rate is increased when chemoreceptors in the carotid artery detect falling blood pH and stimulate the cardio-acceleratory centre. This stimulates the sino-atrial node via sympathetic nerve fibres to increase the rate of electrical excitation of the heart muscle.

- Heart rate is decreased when chemoreceptors in the carotid artery detect rising blood pH and stimulate the cardio-inhibitory centre. This stimulates the sino-atrial node via parasympathetic nerve fibres to decrease the rate of electrical excitation of the heart muscle.

Key Terms

Cerebrum: contains two hemispheres responsible for integrating sensory functions and initiating voluntary motor functions. It is the source of intellectual function in humans, where it is more developed than in other animals.

Cerebellum: part of the hindbrain that coordinates the precision and timing in muscular activity, contributing to equilibrium and posture, and to learning motor skills.

Medulla oblongata: part of the hindbrain that connects the brain to the spinal cord and controls involuntary, autonomic functions.

Autonomic nervous system: the part of the peripheral nervous system that controls automatic functions of the body by the antagonistic activity of the sympathetic and parasympathetic nervous systems.

Cerebral cortex

The brain contains two hemispheres, which are connected to each other by a bundle of nerves called the corpus callosum. The outermost layer (2–3mm) is the cerebral cortex and is highly folded, which increases the area available for processing information, and contains thousands of millions of neurones each with many synaptic connections. The cerebral cortex is responsible for conscious thoughts and actions. The cortex consists of grey matter with many cell bodies, but the inner cerebrum consists of white matter due to the presence of myelin in the myelinated axons.

Each cerebral hemisphere is subdivided into four structural regions called lobes each with distinct functions:

1. Frontal lobe's activity contributes to reasoning, planning, speech and movement, emotions and problem solving.
2. Temporal lobe involved in language, learning and memory.
3. Parietal lobe involved in somatosensory functions which can occur in any part of the body, and taste.
4. Occipital lobe involved in vision.

The cerebral cortex can be subdivided into three discrete functional areas:

1. Sensory areas or cortex, which receive nerve impulses from receptors in the body via the thalamus.
2. Motor areas or cortex, which send nerve impulses to effectors via motor neurones which cross over in the medulla oblongata so the left hemisphere controls the right side of the body.
3. Association area, which makes up most of the cerebral cortex and receives impulses from sensory areas. It associates the information received with previously stored information allowing the information to be interpreted and given meaning. It is also responsible for initiating appropriate responses which are passed to the relevant motor areas.

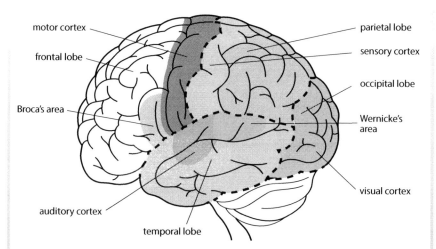

Functional areas of the human brain

The size of the sensory and motor areas associated with different parts of the body varies. Where there are large numbers of sensory neurones, e.g. in tongue, lips and fingers, a large part of the sensory area is dedicated to receiving impulses from these areas. This is shown by the sensory **homunculus**. Similarly, muscles of the face and hands require many motor neurones to control the fine movements, and so a large proportion of the motor area is involved, which is shown by the motor homunculus.

quickfire

54 Which lobe of the cerebral cortex is involved in
a) language, learning and memory, b) vision and c) speech and movement?

Key Term

Homunculus: a drawing of the relationship between the complexity of innervation of different parts of the body and the areas and positions in the cerebral cortex that represent them.

Language and speech

Language and speech involve Wernicke's area, which is an association area that interprets written and spoken language, and Broca's area in the motor area, which innervates (supplies nerves to) the muscles, e.g. intercostal muscles and muscles in mouth, and larynx required to produce sound. Both areas are located in the left hemisphere of the brain and are linked by a bundle of nerve fibres called the arcuate fasciculus.

Neuroscience

A number of non-invasive imaging techniques can be used to reveal brain structure and function:

1. Electroencephalography (EEG) involves the placing of electrodes on the scalp which record general changes in the electrical activity of regions of the brain over time.

2. Computerised tomography (CT) scans combine many X-ray images taken from different angles using computers to produce multiple high resolution images of the brain in cross section.

3. Magnetic resonance imaging (MRI) uses a powerful magnetic field, radio frequency pulses and a computer to produce 3D images which are more detailed than CT scans showing clearly the difference between grey and white matter in the brain.

4. Functional magnetic resonance imaging (fMRI) is a technique for examining activity of brain tissue in real time as opposed to structure. It uses a radio wave pulse (in addition to a magnetic field) which interacts differently with haemoglobin and oxyhaemoglobin showing areas where there is greater oxygen demand suggesting greater brain activity.

5. Positron emission tomography (PET) is a neuroimaging technique which involves the injection of a small amount of radioactive isotope fluorodeoxyglucose, which is taken up into active cells and then emits a positron as it decays. This is detected by the scanner and shows areas where respiration is occurring more. The isotope used has a short half-life and so is quickly eliminated from the body.

quickfire

⑤⑤ Which imaging technique involves the injection of a small amount of radioactive isotope?

Neuroplasticity

Many aspects of the brain remain plastic and therefore can change even into adulthood, making new neuronal connections in response to sensory information, development or damage. Following a stroke or other brain injury, undamaged axons form new pathways with damaged neurones. There are two types of **neuroplasticity**:

1. Synaptic plasticity is the ability of a synapse to alter the amount of neurotransmitter released or the response initiated in the post synaptic neurone. This plays a significant role in learning and memory.

2. Non-synaptic plasticity involves the excitability of an axon changing by altering the voltage-gated channels often as a result of injury.

quickfire

⑤⑥ Name the type of neuroplasticity which involves the excitability of an axon changing by altering the voltage-gated channels.

Neuroplasticity: the ability of the brain to modify its own structure and function following changes within the body or in the external environment.

Synaptic pruning: the elimination of synapses that happens between early childhood and maturity.

Synaptic pruning, which occurs in adolescence, reduces the number of synapses per neurone from around 15,000 per neurone by age 3 to 1000–10,000 per neurone by adulthood in response to interactions with the environment. The result is 'hard-wired' connections allowing quick and accurate transmission of signals.

The development of language in humans requires extensive postnatal experience to produce and decode speech sounds that are the basis of language and so must occur in early life. A language developmental delay can result if hearing is compromised, e.g. as a result of repeated ear infections during the first few years and with deaf children unless they can express themselves by another means, e.g. signing. A well-documented example involving a 'feral' girl deprived of language up to the age of 13 years meant that despite intensive training, she only ever managed basic communication.

Mental illness

Studies have shown that mental illnesses such as depression and schizophrenia are caused in part by genetic factors, but the causes are complex and varied. Many conditions are polygenic, and epigenetics is thought to play a part as many brain functions are accompanied by changes in gene expression at a cellular level.

Abused children are at least 50% more likely than the general population to suffer from serious depression as adults, and find it harder to recover. They are also at significantly higher risk of a range of other conditions including schizophrenia, eating disorders, personality disorders, bipolar disease and general anxiety, as well as being more likely to abuse drugs or alcohol. It is thought that bad experiences during a key developmental period in childhood result in physical changes in the brain which may have an epigenetic component.

Role of cortisol

Cortisol production is controlled by the hippocampus and is released from the adrenal glands into the blood, and functions to increase blood glucose and suppress the immune system. Its production increases during periods of stress:

- The hippocampus sends impulses to the hypothalamus, which releases two hormones, corticotrophin-releasing hormone and arginine vasopressin.
- These hormones stimulate the pituitary gland to release the hormone adrenocorticotrophin into the blood, resulting in cells of the adrenal glands releasing cortisol.
- Cortisol binds to glucocorticoid receptors on the hippocampus, which responds by sending nerve impulses to the hypothalamus, inhibiting further release of cortisol. This is an example of negative feedback.

It appears that adults who suffered traumatic childhoods are constantly over-stressed because they produce too much cortisol, suggesting that the feedback loop is not functioning properly.

Behaviour

Behaviour may be either innate (inborn) which is instinctive, or learned.

1. Innate behaviour is more significant in animals with less complex neural systems as they are less able to modify their behaviour as a result of learning. It includes:

   - Reflexes (see page 66) which are rapid and automatic and protect part of an organism from harm.

   - Kineses are more complex than reflexes and involve the movement of the whole organism, and are non-directional, resulting in a faster movement or a change in direction.

   - Taxes involve the whole organism moving in response to a stimulus, where the direction of the movement is related to the direction of the stimulus either towards or away from it. An example is seen in woodlice which show negative phototaxis by moving away from light.

2. Learned behaviour builds upon and modifies existing knowledge resulting in a relatively permanent change in behaviour or skills.

   - Habituation involves learning to ignore stimuli because they are not followed by either reward or punishment.

   - Imprinting occurs at a very early age in a critical period of brain development in birds and some mammals. Konrad Lorenz noted that the young of birds, and some mammals, respond to the first larger moving object they see, smell, touch or hear. They attach to this object and the attachment is reinforced by rewards, e.g. food.

   - Associative behaviours include classical and operant conditioning, in which animals associate one type of stimulus with a particular response or action:

     - Classical conditioning involves the association between a natural and an artificial stimulus to bring about the same response. Ivan Pavlov conducted experiments with dogs in which he used a 'neutral stimulus' of a bell ringing which the dogs learned to associate with food. The dogs would salivate in response to the bell even in the absence of food.

     - Operant conditioning involves the association between a particular behaviour and a reward or punishment. BF Skinner conducted experiments with mice where they learned to press a lever to receive food (reward) or to stop a loud noise (punishment).

     - Latent (exploratory) learning is not directed to satisfying a need or obtaining a reward. Many animals explore new surroundings and learn information which, at a later stage, can mean the difference between life and death.

     - Insight learning does not result from immediate trial and error learning but may be based on information previously learned by other behavioural activities. Kohler conducted experiments with chimpanzees in the 1920s where they were given food but it was out of reach. Chimpanzees were given sticks and boxes, and eventually they learned to use them to reach the food.

quickfire

57 What is the main difference between taxes and kineses?

quickfire

58 Name the type of learning that involves the association between a particular behaviour and a reward or punishment.

quickfire

59 Name the type of learning that involves the association between a natural and an artificial stimulus to bring about the same response.

- Imitation is an advanced form of social learning that allows learned behaviour patterns to spread rapidly between individuals and to be passed down from generation to generation. It involves the copying of the behaviour of another animal. It can result in differences between social groups, e.g. some chimpanzees crack nuts using stones, whilst others use branches.

Living in social groups

Many species form highly structured social groups called societies, where the behaviour of one can influence others within the group. This social behaviour relies on the ability of animals to communicate with each other by one animal using sign stimuli (signals) which are detected by another, triggering an innate response, e.g. the begging response of a gull chick is triggered by the red spot on its parent's beak. These are often referred to as **fixed action patterns** (FAPs) which are a form of stereotyped behaviour where the sign stimulus activates nerve pathways bringing about coordinated movements without any conscious thought. The response of the individual is dependent upon its motivational state, e.g. a cheetah will only initiate stalking behaviours upon the sight of prey if it is hungry.

Tinbergen conducted experiments with foraging honey bees showing that they showed curiosity for yellow and purple model flowers but only landed to feed if the proper odour was also present. Only then would the bees insert their mouthparts and feed. These behaviour types are more complex than reflexes and can be modified by experience.

Insect social structures

Many insects are social and live in colonies based upon a caste system, e.g. ants, bees and termites. The colonies involve brood care for offspring, overlapping generations in the colony, and a division of labour where individuals cannot perform tasks of individuals in other castes. This leads to high efficiency within the group. Members communicate through touch, visual orientation and pheromones.

A honeybee colony consists of one female (the queen), a few hundred reproductive males (drones) and tens of thousands of sterile female workers. The workers find nectar, clean the hive, care for the young, build new wax combs and defend the colony. Karl von Frisch discovered how honeybees communicate the distance and direction of the food source to other worker bees by the use of a dance performed at the hive entrance: A round dance indicates the source is less than 70m from the hive. A waggle dance indicates that the source is over 70m away and its direction relative to the hive and the sun.

Vertebrate social structure

Social groups in vertebrate animals are based largely upon **dominance hierarchies** where a dominant individual (often an alpha male) is dominant over others. The hierarchies tend to be linear and so have no members of equal rank, e.g. a group of hens sharing a hen-house. This type of hierarchy only exists where animals are able to recognise each other as individuals and are able to learn.

Dominance hierarchy decreases the amount of individual aggression associated with: feeding, mate selection, breeding site selection, and ensures that resources are shared. It also ensures that the fittest male has greater access to females and so produces fitter offspring in a male-dominant hierarchy.

A dominance hierarchy is relatively stable as fighting is a last resort. In red deer prior to mating, there is a series of ritualised actions, e.g. during the mating season, stags roar and walk parallel with each other to assess their opponent's strength. If the weaker stag fails to withdraw, they will butt antlers. If this is prolonged, injuries can arise.

Key Term

Dominance hierarchy: ranking system in an animal society, in which each animal is submissive to animals in higher ranks but dominant over those in lower ranks.

Territorial and courtship behaviour

Most animals occupy an area called the home range, but only a few defend a territory against other members of the same species. Territories are marked by scent, visual signposts, e.g. faeces, and sounds or calls.

Courtship is used to attract a mate and allows sexually receptive individuals of the same species to be recognised. Elaborate courtship routines have evolved and are innate ensuring intraspecific mating, e.g. sticklebacks. If an egg-carrying female approaches, the male starts to zigzag, which entices her to swim closer, which in turn stimulates him to swim to the nest and stick his snout inside. This action stimulates the female to wriggle into the tunnel. The male then nuzzles her tail, which stimulates her to spawn, after which she swims out the other end of the nest. The male then enters and deposits sperm on the eggs.

Many species show sexual dimorphism where the males look different from the females, e.g. peahens and peacocks. The peacock tail is a large encumbrance and makes the male more susceptible to predation, but it does indicate his genetic fitness to the female.

There are two main theories behind the driving force of sexual selection:

1. Intra-sexual selection or male–male combat. Some species, e.g. African lions, and southern elephant seals, have males which are much bigger than the females. Males fight for access to the females and so sexual selection has favoured the evolution of larger more aggressive males.

2. Inter-sexual selection or female choice, in which the physical attractiveness model and the male handicap model drive a male's reproductive success.

quickfire

(60) What is the difference between a home range and a territory?

≫ Pointer

With sexual dimorphism, sexual selection and natural selection will be working against each other: Sexual selection will make a characteristic more conspicuous, helping to attract a mate, but this will be costly to the male and will make it more conspicuous to predators.

Unit 4 Summary

4.1 Sexual reproduction in humans

- Gametogenesis is the production of gametes through a series of mitotic and meiotic divisions in the testis and ovaries.
- The menstrual cycle is controlled by gonadotrophic hormones from the anterior pituitary and hormones from the ovary itself.
- Spermatozoa and secondary oocytes are adapted for fertilisation.
- Following fertilisation, a blastocyst forms which implants in the lining of the endometrium leading to pregnancy.

4.2 Sexual reproduction in plants

- Pollination is the transfer of pollen from the anther of one flower to the mature stigma of another flower of the same species.
- Flowers are adapted to transfer pollen from one flower to another, by wind or by insects.
- Plants have evolved different methods for the successful dispersal of their seeds.
- Germination requires optimal conditions including, water, warmth and oxygen.

4.3 Inheritance

- A gene is a sequence of DNA on a chromosome normally coding for a specific polypeptide, which occupies a specific position or locus.
- In co-dominance both alleles involved are dominant and therefore both are expressed equally.
- Autosomal linkage occurs when two different genes are found on the same chromosome and therefore cannot segregate independently.
- Sex linkage is when a gene is carried by a sex chromosome so that a characteristic it encodes is seen predominately in one sex.
- A mutation is a change in the amount, arrangement or structure of the DNA in an organism.
- Duchenne muscular dystrophy (DMD), like haemophilia, is caused by an X-linked recessive allele, but involves the gene which codes for dystrophin, a component of a glycoprotein that stabilises the cell membranes of muscle fibres.

4.4 Variation and evolution

- Variation can be either continuous or discontinuous.
- The *t*-test can be used to see if the means of two sets of data are significantly different.
- Environmental influences affect the way a genotype is expressed and results in different phenotypes.
- Competition may be either intraspecific or interspecific.
- The Hardy-Weinberg principle states that in ideal conditions the allele and genotype frequencies in a population are constant.
- Changes to environmental conditions bring new selection pressures through competition, predation or disease, which results in a change in the allele frequency.
- Speciation is the evolution of new species from existing ones.

4.5 Application of reproduction and genetics

- Many organisms have had their genomes sequenced.
- The polymerase chain reaction (PCR) can be used to copy a large number of specific fragments of DNA rapidly.
- Gel electrophoresis is a method of separating DNA fragments according to size.
- Genetic engineering allows genes to be manipulated, altered or transferred from one organism or species to another, making a genetically modified organism (GMO).
- Restriction enzymes are bacterial enzymes that cut up any foreign DNA which enters a cell.
- DNA ligase is a bacterial enzyme that joins sugar-phosphate backbones of two molecules of DNA together.
- Reverse transcriptase is an enzyme that produces DNA from a RNA template.
- Bacteria can be genetically engineered to produce medicinal products, e.g. insulin.
- Crops can be genetically engineered to be resistant to disease and pests.
- Gene therapy can be used to treat a number of genetic diseases including cystic fibrosis and DMD.

Option A Immunology and disease

- A disease refers to an illness of people, animals or plants which is caused by an infection or failure of health.
- Cholera is caused by the Gram-negative bacterium *Vibrio cholera.*
- Tuberculosis (TB) is caused by the bacillus bacterium *Mycobacterium tuberculosis.*
- Smallpox is caused by the virus *Variola major.*
- Influenza virus has three subgroups which contain viruses with different antigenic types. Influenza virus attacks the mucous membranes in the upper respiratory tract.
- Malaria is caused by a protoctistan parasite *Plasmodium.* The cycle of red blood cells bursting, releasing merozoites which go on to infect other cells repeats every few days and gives rise to recurring fevers.
- Viruses are intracellular parasites that use the host cell's metabolic pathways to reproduce, they produce pathogenic effects to the host.

- The structure of the bacterial cell wall greatly influences the type of antibiotic that will be effective.
- Gram-positive bacteria have thicker cell walls than Gram-negative bacteria, containing peptidoglycan with polysaccharide molecules cross-linked to amino acid side chains. Gram-negative bacteria have thinner but more complex cell walls.
- Antibiotic resistance is the ability of a microorganism to withstand the effects of an antibiotic.
- There are two components to the immune response: The humoral response, which results in the production of antibodies and the B-lymphocytes, and cell-mediated response which involves activating B- and T-lymphocytes and phagocytic cells.
- The primary response follows the first exposure to a given antigen, secondary response follows exposure to the same antigen, which results in a much more rapid response.
- Passive immunity occurs when the body receives antibodies, either naturally or artificially.
- Active immunity occurs when the body produces its own antibodies in response to antigens being present.

Option B Human musculoskeletal anatomy

- The three main tissues involved in the musculoskeletal system are cartilage, bone and skeletal muscle.
- Cartilage is a hard and flexible connective tissue that permits movement of structures, e.g ribcage, whilst its rigidity provides support to structures, e.g. trachea.
- Bone provides structural support, provides muscle attachment sites, and is also involved in mineral regulation.
- Rickets results from the inadequate deposition of minerals in the growing bones of children as a result of a lack of the fat soluble vitamin D or calcium in the diet.

- Osteoporosis is the abnormal loss of density in spongy and compact bone, which results in increased risk of fractures.
- Skeletal muscle is made up of muscle fibres which are long thin cells containing many nuclei.
- The sliding filament theory is the theory of muscle contraction in which thin, actin filaments slide between thick myosin filaments in response to a nervous impulse mediated by the T system.
- Slow twitch which contract slowly but with less force than fast twitch fibres. Fast twitch which contract quickly with more force. The proportions of slow and fast twitch fibres can be altered by training.

Option B cont ...

- Fractures can be caused by high impact or stresses where the force on the bone exceeds its strength, e.g. through trauma.

- The shape of the vertebrae and angle of the facet joints and spinous processes vary down the spine which allows for varying degrees of movement.

- A joint occurs where bones meet. Joints are classified according to the type of movement possible.

- Antagonistic muscles work in pairs to coordinate movement at joints where a flexor muscle allows for bending and an extensor muscle for straightening.

- When a lever is at equilibrium, $F_1 \times d_1 = F_2 \times d_2$ where F_1 is the force exerted by the load and F_2 is the force exerted by the effort, and d represents the distances of the load and effort from the fulcrum.

- Osteoarthritis is the commonest joint disease and is degenerative, and involves the breakdown of articular cartilage faster than it can be replaced.

- Rheumatoid arthritis is an autoimmune disorder that attacks bone and cartilage at joints resulting in severe inflammation.

Option C Neurobiology and behaviour

- The brain is responsible for coordinating responses to sensory stimuli. It is surrounded by three membranes called meninges.

- The brain has three regions: forebrain, midbrain and hindbrain.

- The cerebrum controls voluntary behaviour and learning, medulla oblongata controls heart rate and ventilation. The cerebellum is involved with the maintenance of posture, and the hypothalamus regulates body temperature and blood glucose concentration.

- Heart rate is regulated by changes in blood pH which happens during exercise when carbon dioxide builds up in the blood.

- Each cerebral hemisphere is subdivided into four structural regions called lobes:

  - The frontal lobe involved in reasoning, planning, part of speech and movement, emotions and problem solving.

  - The temporal lobe involved in language, learning and memory.

  - The parietal lobe involved in somatosensory functions and taste.

  - The occipital lobe involved in vision.

- Language and speech involve Wernicke's area, which is an association area that interprets written and spoken language, and Broca's area in the motor area, which innervates (supplies nerves to) the muscles required to produce sound.

- A number of non-invasive imaging techniques can be used to reveal brain structure and function including EEG, CT, MRI, fMRI and PET scans.

- Neuroplasticity is the ability of the brain to modify its own structure and function following changes within the body or in the external environment.

- Studies have shown that the risk of developing mental illnesses such as depression and schizophrenia are caused in part by genetic factors.

- Behaviour may be either innate (inborn) which is instinctive or learned.

- Social structures have evolved in animals which can involve cooperative rearing of young, foraging or hunting, defence and learning.

Exam practice and technique

Aims and Objectives

The WJEC A Level in Biology aims to encourage learners to:

- Develop essential knowledge and understanding of different areas of biology and how they relate to each other.
- Develop and demonstrate a deep appreciation of the skills, knowledge and understanding of scientific methods used within biology.
- Develop competence and confidence in a variety of practical, mathematical and problem solving skills.
- Develop their interest in and enthusiasm for biology, including developing an interest in further study and careers associated with the subject.
- Understand how society makes decisions about biological issues and how biology contributes to the success of the economy and society.

Types of exam question

There are **two** main types of question in the exam:

1. Short and longer-answer structured questions

The majority of questions fall into this category. These questions may require description, explanation, application, and/or evaluation, and are generally worth 6–10 marks. Application questions could require you to use your knowledge in an unfamiliar context or to explain experimental data. The questions are broken down into smaller parts, e.g. (a), (b), (c), etc., which can include some 1-mark name or state questions, but most will require description, explanation or evaluation for 2–5 marks. You could also be asked to complete a table, label or draw a diagram, plot a graph, or perform a mathematical calculation.

Some examples requiring 'name', 'state' or 'define':

- Define the term biodiversity. (1 mark)
- State the term used to describe the transfer of energy between consumers. (1 mark)
- Name the cells shown that are undergoing meiosis. (1 mark)
- Identify hormone A shown in the graph. (1 mark)

Some examples requiring mathematical calculation:

- The magnification of the image above is × 32,500. Calculate the actual width of the organelle in micrometres between points A and B. (? marks)
- Using the graph, calculate the initial rate of reaction for the enzyme. (2 marks)
- Calculate the percentage energy lost through respiration by secondary consumers. (2 marks)
- Use the Hardy–Weinberg formula to estimate the number of individuals in a population of 1000 that would be carriers of the condition. (4 marks)
- Calculate χ^2 for the results of the cross shown. (3 marks)

Some examples requiring description:

- Describe how biodiversity loss could be delayed. (1 mark)
- Describe how a sweep net could be used to estimate the diversity index of insects at the base of a hedge. (3 marks)

Some examples requiring explanation:

- Suggest one limitation of the method used, and explain how this could have affected the validity of the conclusion drawn. (2 marks)
- Explain why there must be three bases in each codon to assemble the correct amino acid. (2 marks)
- Explain the term planetary boundary. (2 marks)
- Explain why it is important when using a biosensor to measure urea concentration to maintain a constant temperature and pH. (2 marks)
- Explain how the structures of cellulose and chitin are different from that of starch. (2 marks)

Some examples requiring application:

- Suggest the function of NAD in the series of reactions shown. (1 mark)
- A drug has been shown to block the initiation of S phase in mitosis. Suggest why this could be used to treat cancer. (3 marks)
- Use the information provided to explain why sodium benzoate would affect the accuracy of the biosensor. (5 marks)

Some examples requiring evaluation:

- Describe how you could improve your confidence in your conclusion. (2 marks)
- Analyse the data in the table and draw alternative conclusions. Explain how you reached these conclusions. (3 marks)
- Evaluate the strength of their evidence and hence the validity of their conclusion. (4 marks)

2. Extended response questions

One question in each component exam contains an extended response question worth 9 marks. The quality of your extended response (QER) will be assessed in this question. You will be awarded marks based upon a series of descriptors: to gain the top marks it is important to give a full and detailed account, including a detailed explanation. You should use scientific terminology and vocabulary accurately, including accurate spelling and use of grammar and include only relevant information. It is a good idea to do a brief plan before you start, to organise your thoughts: You should cross this out once you have finished. We will look at some examples later.

Command or action words

These tell you what you need to do. Examples include:

Analyse means to examine the structure of data, graphs or information. A good tip is to look for trends and patterns, and maximum and minimum values.

Calculate is to determine the amount of something mathematically. It is really important to show your working (if you don't get the correct answer you can still pick up marks for your working).

Choose means to select from a range of alternatives.

Compare involves you identifying similarities and differences between two things. It is important when detailing similarities and differences that you talk about both: a good idea is to make two statements, linked with the word '*whereas*'.

Complete means to add the required information.

Consider is to review information and make a decision.

Describe means give an account of what something is like. If you have to describe the trend in some data or in a graph then give values, e.g. is there a peak, trough or plateau?

Discuss involves presenting the key points.

Distinguish involves you identifying differences between two things.

Draw is to produce a diagram of something.

Estimate is to roughly calculate or judge the value of something.

Evaluate involves making a judgment from available data, conclusion or method, and proposing a balanced argument with evidence to support it.

Explain means give an account and use your biological knowledge to give reasons why.

Identify is to recognise something and be able to say what it is.

Justify is about you providing an argument in favour of something; for example, you could be asked if the data supports a conclusion: You should then give reasons why the data support the conclusion given.

Label is to provide names or information on a table, diagram or graph.

Name is to identify using a recognised technical term. Often a one-word answer.

Outline is to set out the main characteristics.

State means give a brief explanation.

Suggest involves you providing a sensible idea. It is not straight recall, but more about applying your knowledge.

General exam tips

Always read the question carefully: Read the question twice! It is easy to provide the wrong answer if you don't give what the question is asking for. Information provided in the question is there to help you to answer it. The skill is to identify which information is relevant to the part of the question you are answering. The wording has been discussed at length by examiners to ensure that it is as clear as possible.

One of the most frequent comments that examiners make is that students' answers often lack detail or don't explain fully: We will look at some examples of these later.

Look at the number of marks available. A good rule is to make *at least* one different point for each mark available. So make five different points if you can for a four-mark question to be safe. Make sure that you keep checking that you are actually answering the question that has been asked – it is easy to drift off topic! If a diagram helps, include it: But make sure it is fully annotated.

Timing

There is one written examination paper for each unit, each lasting 2 hours. Each examination is out of 90 marks and contributes 25% of the final grade. In Unit 4, Section B contains a choice of one question from three worth 20 marks: You should answer the question from the topic that you have studied.

The options are: Immunology and Disease, Human Musculoskeletal Anatomy and Neurobiology and Behaviour.

The number of marks available gives you a sense of how much time you should spend on each exam question, and a good rule is about one mark per minute. Don't forget that this timing is not just about writing but you should spend time thinking, and for the extended answer some planning, too.

Unit 5 consists of a practical examination worth 50 marks which contributes 10% of the final grade. The unit comprises of an experimental task (20 marks) and a practical analysis task (30 marks).

Synoptic assessment

To prepare for the A2 exams you will need to cover both AS and A2 work. Some questions set in A2 units will require you to draw together different areas of knowledge from across AS and A2 units, examples of this could include:

- Enzyme action in respiration and photosynthesis.
- DNA structure and genetic code in application of genetics.
- Cell structure in microbiology.
- Classification in variation and evolution.
- Meiosis in sexual reproduction and variation.
- Cell membranes and transport in the nervous system.

Assessment objectives

Examination questions are written to reflect the assessment objectives (AOs) as laid out in the specification. The three main skills that you must develop are:

AO1: Demonstrate knowledge and understanding of scientific ideas, processes, techniques and procedures.

AO2: Apply knowledge and understanding of scientific ideas, processes, techniques and procedures.

AO3: Analyse, interpret and evaluate scientific information, ideas and evidence, including in relation to issues.

In written examinations you will also be assessed on your:

- Mathematical skills (minimum of 10% of available marks).
- Practical skills (minimum of 15% of available marks).
- Ability to select, organise and communicate information and ideas coherently using appropriate scientific conventions and vocabulary.

In any one question, you are likely to be assessed on all skills to some degree. It is important to remember that only about a third of the marks are for direct recall of facts. You will need to apply your knowledge, too. If this is something you find hard, practise as many past paper questions as you can. Examples can come up in slightly different forms from one year to another.

Your practical skills will be developed during class time sessions and will be assessed in the examination papers. These could include:

- Plotting graphs.
- Identifying controlled variables and suggesting appropriate control experiments.
- Analysing data and drawing conclusions.
- Evaluating methods and procedures and suggesting improvements.

Drawing graphs

Full marks are rarely awarded for graphs.

Common errors include:

- Incorrect labels on axes.
- Missing units.
- Sloppy plotting of points.
- Failing to join plots accurately.
- No value for origin on horizontal axis.
- Horizontal axis is non-linear, i.e. gaps are unequal.

Understanding AO1: Demonstrate knowledge and understanding

You will need to demonstrate knowledge and understanding of scientific ideas, processes, techniques and procedures.

Approximately 27% of the available marks set on the A2 exam papers are for recall of knowledge and understanding.

Common command words used here are: state, name, describe, explain.

This involves recall of ideas, processes, techniques and procedures detailed in the specification. This is content you should know.

A good answer is one that uses detailed biological terminology accurately, and has both clarity and coherence.

If you were asked to describe and explain how electrophoresis produced the results seen in a gel, you might write:

'DNA moves towards the positive electrode through the gel. Smaller fragments move further.'

This is a basic answer.

A good answer needs to be more detailed. For example,

'DNA is attracted to the positive electrode due to the negative charge on its phosphate groups.

Smaller fragments find it easier to migrate through the pores in the gel and so travel further than larger fragments in the same time. The size of fragment can be estimated by running a DNA ladder which contains fragments of known size alongside the sample.'

Understanding AO2: Applying knowledge and understanding

You will need to apply knowledge and understanding of scientific ideas, processes, techniques and procedures:

- In a theoretical context.
- In a practical context.
- When handling qualitative data (this is data with no numerical value, e.g. a colour change).
- When handling quantitative data (this is data with a numerical value, e.g. mass/g).

45% of the available marks on the A2 exam papers are for application of knowledge and understanding.

Common command words used here are: describe (if it's unfamiliar data or diagrams), explain and suggest.

AO2 tests applying ideas, processes, techniques and procedures detailed in the specification to unfamiliar situations including using mathematical calculations and interpreting the results of statistical tests.

If you were asked to describe the effects of a weed-killer on non-cyclic photophosphorylation explaining why cyclic photophosphorylation was unaffected given the information that the weed-killer blocks electron flow from Photosystem II to the electron carrier, you might write:

'It stops electrons moving out of Photosystem II into the electron carrier so electrons can't pass to Photosystem I.'

This is an incomplete answer and does not explain why cyclic photophosphorylation is unaffected.

A good answer would say:

'The weed-killer stops electrons from Photosystem II being moved to Photosystem I which prevents the reduction of NADP to reduced NADP. Photolysis of water cannot occur. Cyclic photophosphorylation is not stopped because the electrons are still able to pass from Photosystem I and return back to Photosystem I.'

Describing data

It is important to describe accurately what you see, and to quote data in your answer.

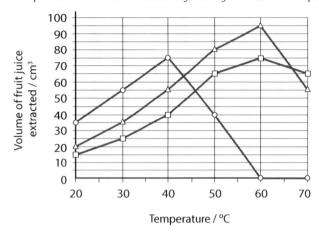

If you were asked to compare the volume of juice produced when using enzymes bound to the gel membrane surface compared with the enzyme immobilised inside the beads, you might write:

'The volume of juice extracted increases with temperature up to the optimum temperature of 60°C in both enzymes. Above this, the volume of juice decreases.'

This is a basic answer.

A good answer needs to be both accurate and detailed. For example,

'Increasing temperature causes the volume of fruit juice extracted to increase up to 60°C. The volume of juice collected is higher up to 60°C with the enzyme bound to the gel membrane, peaking at 95 cm³ compared with 75 cm³ for the enzyme immobilised inside the beads. Above 60°C the volume of fruit juice extracted decreases, but this is more noticeable for the enzymes bound to the gel membrane surface which decrease by 40 cm³ compared with just 10 cm³ for the enzyme immobilised inside the beads.'

If you were also asked to explain the results, a basic answer would include reference to *'increased kinetic energy up to 60°C, and denaturing enzymes above 60°C'.* A good answer is one that uses detailed biological terminology accurately, and has clarity and coherence. A good answer would also include reference to *'increased number of enzyme-substrate complexes forming up to 60°C and would include that above 60°C, hydrogen bonds break resulting in the active site changing shape so fewer enzyme-substrate complexes could form.'*

Mathematical requirements

A minimum of 10% of marks across the whole qualification will involve mathematical content. Some of the mathematical content requires the use of a calculator, which is allowed in the exam. The specification states that calculations of the mean, median, mode and range may be required, as well as percentages, fractions and ratios. The additional requirements included at A level **are shown in bold** on page 165.

You will be required to process and analyse data using appropriate mathematical skills. This could involve considering margins of error, accuracy and precision of data.

Concepts	Tick here when you are confident you understand this concept
Arithmetic and numerical computation	
Convert between units, e.g. mm^3 to cm^3	
Use an appropriate number of decimal places in calculations, e.g. for a mean	
Use ratios, fractions and percentages, e.g. calculate percentage yields, surface area to volume ratio	
Estimate results	
Use calculators to find and use power, exponential and logarithmic functions, e.g. estimate the number of bacteria grown over a certain length of time	
Handling data	
Use an appropriate number of significant figures	
Find arithmetic means	
Construct and interpret frequency tables and diagrams, bar charts and histograms	
Understand the principles of sampling as applied to scientific data, e.g. use Simpson's Diversity Index to calculate the biodiversity of a habitat	
Understand the terms mean, median and mode, e.g. calculate or compare the mean, median and mode of a set of data, e.g. height/mass/size of a group of organisms	
Use a scatter diagram to identify a correlation between two variables, e.g. the effect of lifestyle factors on health	
Make order of magnitude calculations, e.g. use and manipulate the magnification formula : magnification = size of image / size of real object	
Understand measures of dispersion, including standard deviation and range	
Identify uncertainties in measurements and use simple techniques to determine uncertainty when data are combined, e.g. calculate percentage error where there are uncertainties in measurement	
Algebra	
Understand and use the symbols: $=, <, <<, >>, >, \propto, \sim$.	
Rearrange an equation	
Substitute numerical values into algebraic equations	
Solve algebraic equations, e.g. solve equations in a biological context, e.g. cardiac output = stroke volume × heart rate	
Use a logarithmic scale in the context of microbiology, e.g. growth rate of a microorganism such as yeast	
Graphs	
Plot two variables from experimental or other data, e.g. select an appropriate format for presenting data	
Understand that $y = mx + c$ represents a linear relationship	
Determine the intercept of a graph, e.g. read off an intercept point from a graph, e.g. compensation point in plants	
Calculate rate of change from a graph showing a linear relationship, e.g. calculate a rate from a graph, e.g. rate of transpiration	
Draw and use the slope of a tangent to a curve as a measure of rate of change	
Geometry and trigonometry	
Calculate the circumferences, surface areas and volumes of regular shapes, e.g. calculate the surface area or volume of a cell	

Understanding AO3: Analysing, interpreting and evaluating scientific information

This is the last, and most difficult skill. You will need to analyse, interpret and evaluate scientific information, ideas and evidence, to:

- Make judgements and reach conclusions.
- Develop and refine practical design and procedures.

Approximately 28% of the available marks on the A2 exam papers are for analysing, interpreting and evaluating scientific information.

Common command words used here are: evaluate, suggest, justify and analyse.

This could involve:

- Commenting on experimental design and evaluating scientific methods.
- Evaluating results and drawing conclusions with reference to measurement, uncertainties and errors.

What is accuracy?

Accuracy relates to the apparatus used: How precise is it? What is the percentage error? For example, a 5ml measuring cylinder is accurate to +/− 0.1ml so measuring 5ml could yield 4.9–5.1ml. Measuring the same volume in a 25ml measuring cylinder which is accurate to +/−1ml would yield 4–6 ml.

Calculating % error

It's a simple equation: error/initial value ×100. For example, in the 25ml measuring cylinder the accuracy +/− 1ml so the error is 1/25 × 100 = 4%, whereas in the 5ml cylinder the accuracy is +/− 0.1ml so the error is 0.1/5 × 100 = 2%. Therefore, for measuring 5ml it is best to use the smaller cylinder as the percentage error is lowest.

What is reliability?

Reliability relates to your repeats. In other words, if you repeat the experiment three times and the values obtained are very similar, then it indicates that your individual readings are reliable. You can increase reliability by ensuring that all variables that could influence the experiment are controlled, and that the method is consistent, and by performing a number of repeats and then calculating a mean.

Describing improvements

If you were asked to describe what improvements could be made to improve the reliability of the results obtained from an experiment extracting apple juice, you would need to look closely at the method and apparatus used.

Q: Pectin is a structural polysaccharide found in the cell walls of plant cells and in the middle lamella between cells, where it helps to bind cells together. Pectinases are enzymes that are routinely used in industry to increase the volume and clarity of fruit juice extracted from apples. The enzyme is immobilised onto the surface of a gel membrane which is then placed inside a column. Apple pulp is added at the top, and juice is collected at the bottom. The process is shown in the diagram.

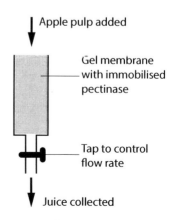

Apparatus used

You might write:

'I would make sure that the same mass of apples is added, and that they were the same age.'

This is a basic answer.

A good answer needs to be both accurate and detailed. For example,

'I would make sure that the same mass of apples is added, for example 100g, and that they were the same age, e.g. 1 week old. I would also control the temperature at an optimum for the pectinases involved, e.g. 30°C.'

Look at the following example:

A student carried out an experiment to investigate the effect of temperature on respiration in yeast cells. 1g of dried yeast was added to 25 cm³ of a 5% glucose solution and after 10 minutes incubation at 15°C, 1cm³ of 5% TTC solution was added. TTC is an artificial hydrogen acceptor which turns from colourless to red in the presence of hydrogen atoms which are released during respiration. The time taken for the yeast solution to turn red was recorded. The experiment was repeated at 30 °C and 45 °C and the time taken for the yeast suspension to turn red was recorded below.

Temperature (°C)	Time taken for the yeast suspension to turn red (s)			
	Trial 1	Trial 2	Trial 3	Mean (nearest whole second)
15	450	427	466	448
30	322	299	367	329
45	170	99	215	161

Q: What conclusions could be drawn from this experiment regarding the effect of temperature on respiration in yeast?

You might write:

'Increasing temperature decreases the time taken for the yeast suspension to turn red, indicating that respiration is occuring more quickly.'

A good answer needs to be both accurate and detailed. For example,

'Increasing the temperature increases the rate of respiration in the yeast, so dehydrogenase enzymes remove hydrogen atoms from triose phosphate more quickly. This is due to increased kinetic energy of the dehydrogenase enzymes and triose phosphate substrate molecules at higher temperatures. As more hydrogen atoms are released more quickly, so TTC is reduced more quickly turning the yeast red in a shorter time.'

If asked to comment on the validity of your conclusion, you might write:

'It was difficult to determine when the solutions turned red, making it difficult to know when to stop timing the reactions.'

A good answer would be more detailed, for example,

'The results at 45°C are very variable and range from 99 to 215 seconds. It is difficult to reach a conclusion about the effect of temperature on respiration in yeast as only three temperatures were investigated. Another major difficulty would be in determining the end point of the reaction, as no standard red colour or colorimeter was used.'

As part of this skill you could also be asked to identify the independent, dependent and controlled variables in an investigation. Remember:

- The independent variable is the one I change.
- The dependent variable is the one I measure.
- Controlled variables are variables which affect the reaction being investigated, and must be kept constant.

Questions and answers

This part of the guide looks at actual student answers to questions. There is a selection of questions covering a wide variety of topics. In each case there are two answers given; one from a student (Lucie) who achieved a high grade and one from a student who achieved a lower grade (Ceri). We suggest that you compare the answers of the two candidates carefully; make sure you understand why one answer is better than the other. In this way you will improve your approach to answering questions. Examination scripts are graded on the performance of the candidate across the whole paper and not on individual questions; examiners see many examples of good answers in otherwise low scoring scripts. The moral of this is that good examination technique can boost the grades of candidates at all levels.

Unit 3

Photosynthesis

An experiment was carried out using algae in Calvin's lollipop flask. At regular intervals over one hour, samples were removed into a tube which contained hot methanol. The products were identified and their masses measured using mass spectroscopy. The experiment was carried out once using 0.04% hydrogen carbonate and repeated using 0.008%. The relative masses of glycerate-3-phosphate (GP), triose phosphate (TP) and ribulose bisphosphate (RuBP) are shown on the graph below.

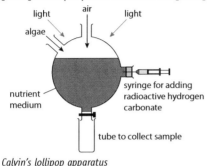

Calvin's lollipop apparatus

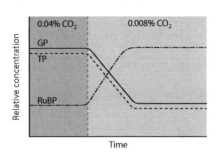

Results graph

a) Suggest why samples were collected in a tube containing hot methanol, explaining why results would be less reliable if this were not done. (2)

b) Describe and explain the effect of decreasing hydrogen carbonate concentration on the relative concentrations of GP, TP and RuBP. (5)

Lucie's answer

a) Contains hot methanol to denature enzymes to prevent any further reactions ✓ If this was not done further products could be made, e.g. GP could be converted to TP. ✓

b) Decreasing concentration of hydrogen carbonate means that less carbon dioxide is available to join with RuBP to produce GP, so RuBP accumulates. ✓ ① and therefore less GP can be produced. Any GP present is being converted into TP and so GP falls. ✓ TP concentration falls because of a reducing GP concentration and any TP present will be converted to carbohydrate, so TP is used up. ✓

Examiner commentary

Good answer from Lucie. ① The answer could have included that RuBisCO is responsible for fixing carbon dioxide and RuBP and therefore fewer successful enzyme-substrate collisions occur due to the lower concentration of substrate.

Lucie achieves 5/7 marks

Ceri's answer

a) To kill algae to stop reactions ✗ ①

b) Decreasing concentration of hydrogen carbonate causes RuBP concentration to increase. ✗ ② GP decreases when hydrogen carbonate decreases because GP not being made from RuBP, but is being converted into TP. ✓ ③

Examiner commentary

① Ceri fails to say that enzyme reactions will be stopped: This is important as GP could be further converted to TP unless enzymes are denatured.

② This is purely descriptive: There is no link made between hydrogen carbonate and carbon dioxide concentrations, and there is no mention of the enzyme involved RuBisCO or enzyme kinetics.

③ Ceri fails to mention the reasons for a decrease in TP: That TP is being converted into carbohydrate.

Ceri achieves 1/7 marks

Exam tip

It is important that you read the question carefully and give as detailed response as you can. You must name enzymes involved and explain the results in terms of enzyme kinetics, that were covered in AS.

Respiration

A sample of calf liver was homogenised in an ice-cold isotonic buffer solution, and then centrifuged at high speed to separate the organelles. The supernatant and a sample of mitochondria were incubated with glucose and the products identified. The experiment was repeated with both samples using pyruvate instead of glucose. The results are shown in the table.

Sample	Incubated with glucose		Incubated with pyruvate	
	CO_2 produced	Lactate produced	CO_2 produced	Lactate produced
Mitochondria	✗	✗	✓	✗
Supernatant	✗	✓	✗	✗

a) Explain why the buffer used was isotonic. (1)
b) Explain the results observed. (5)

Lucie's answer

a) Isotonic buffer prevents lysis of mitochondria. ✓

b) Mitochondria are unable to metabolise glucose which is why neither carbon dioxide or lactate is produced. ✓ ① When mitochondria are incubated with pyruvate, carbon dioxide is produced, because pyruvate can diffuse into the mitochondria and carbon dioxide is produced as a result of Link reaction, and Krebs cycle that occur within the mitochondrial matrix. ✓ The supernatant contains the enzymes present in the cytoplasm of the liver cells and so glycolysis occurs, producing lactate from glucose by anaerobic respiration. ✓ No lactate or carbon dioxide are produced when supernatant is incubated with pyruvate because the link reaction and Krebs cycle do not occur in the cytoplasm. ✓

Examiner commentary

Good answer from Lucie. ① The answer could have included that glycolysis does not occur in the mitochondria and therefore will not have the enzymes necessary to break down glucose.

Lucie achieves 5/6 marks

Ceri's answer

a) Buffer maintains a constant pH. ✗ ①

b) Mitochondria cannot produce carbon dioxide or lactate. ✗ ② Mitochondria release carbon dioxide when incubated with pyruvate, because pyruvate is hydrolysed during Link, and Krebs cycle. ✓ ③ The supernatant is made from cell cytoplasm which is where glycolysis occurs, so lactate is produced by anaerobic respiration. ✓ ④.

Examiner commentary

① Ceri has confused isotonic buffer with pH buffer.

② This is purely descriptive, there is no explanation.

③ Ceri mentions the reactions but could have included where they take place, i.e. mitochondrial matrix.

④ Ceri did not explain why carbon dioxide and lactate are not produced when the supernatant is incubated with pyruvate, i.e. because the link and Krebs cycle do not occur in the cytoplasm.

Ceri achieves 2/6 marks

Exam tip

It is important that you refer to stages in respiration and state exactly where they occur.

Microbiology

Following an outbreak of food poisoning, samples of food were tested using the Gram stain and the bacteria were found to be red in colour. Using the viable cell count method, 1 cm³ of food sample was diluted by adding to 9 cm³ of sterile water using aseptic technique. The sample was mixed, and dilutions repeated. 0.1 cm³ of each dilution was then spread onto a sterile agar plate and the plates incubated at 37°C for 24 hours. The results are shown below. TMC indicates too many to count.

Dilution factor	Number of colonies grown
10^{-1}	TMC
10^{-2}	TMC
10^{-3}	871
10^{-4}	85
10^{-5}	8

a) Suggest *two* reasons why 37°C was chosen rather than 25°C to incubate the plates. (2)

b) Identify which dilution factor should be used, and calculate the number of live bacteria per cm³ in the original food sample. (3)

c) Explain why penicillin would not be an appropriate antibiotic to use to treat the patients. (3)

Lucie's answer

a) Bacteria would grow faster at 37°C, ✓ and you would favour growth of human pathogens. ✓

b) 10^{-4} because 10^{-3} contains too many colonies to count accurately, and 10^{-5} too few. ✓
$85 \times 10,000 \times 10 = 8.5 \times 10^{6}$. ✓✓

c) The bacteria stained red so must be Gram negative. ✓ Penicillin is ineffective against Gram-negative bacteria. ✓ ①

Examiner commentary

Good answer from Lucie, with clear working shown for the calculation and use of standard form. ① Lucie would need to include more detail as to why it is ineffective, i.e. because the outer lipopolysaccharide layer prevents the penicillin from reaching the peptidoglycan layer.

Lucie achieves 7/8 marks

Ceri's answer

a) Bacteria grow well at this temperature. ✗ ①

b) $85 \times 10,000 = 850,000$ ✓ ②

c) Penicillin only works against Gram-positive bacteria which would stain purple, not red. ✓ ③

Examiner commentary

① Ceri should include a comparative statement, i.e. that 37°C would result in faster growth than 25°C.

② Ceri should include an explanation why the plate chosen was used. Some credit can be given for working, but Ceri forgot that only 0.1 cm³ was spread, which dilutes the initial suspension a further ten times.

③ Ceri needs to include why penicillin is ineffective.

Ceri achieves 2/8 marks

Exam tip

Always show your working in mathematical calculations as some credit may be given even with an incorrect answer.

Population size and ecosystems

The diagram shows the energy flow through a woodland ecosystem. The photosynthetic efficiency per year, represents the proportion of light energy that is available to plants which is converted (fixed) into chemical energy.

a) Use the information provided to calculate the photosynthetic efficiency of the producers expressed as a percentage to 2 decimal places. **(2)**

b) Calculate the energy lost from primary consumers to decomposers and detritivores (A) per day, to 2 decimal places. **(1)**

c) Distinguish between gross and net primary productivity. **(2)**

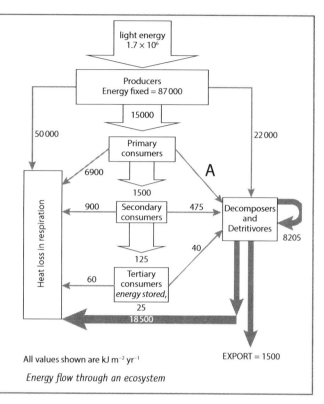

All values shown are kJ m⁻² yr⁻¹

Energy flow through an ecosystem

Lucie's answer

a) $\frac{87,000}{1,700,000} \times 100 = 5.12\ \%.$ ✓

b) $15,000 - 6900 - 1500 = 6,600$ ✓

$\frac{6,600}{365} = 18.08\ \text{kJ m}^{-2}\text{ day}^{-1}.$ ✓

c) GPP represents the rate at which producers convert light energy into chemical energy ✓ whereas NPP represents the rate at which energy is converted into biomass which is available to the next trophic level. ✓

Examiner commentary

Text book answer.

Lucie achieves 5/5 marks

Ceri's answer

a) $\frac{87,000}{1,700,000} \times 100$ ✓ $= 5.11\ \%.$ ✗ ①

b) $15,000 - 6900 - 1500 = 6,600$ ✓ ②

c) $GPP = NPP - R$ ✗ ③

Examiner commentary

① Ceri must to round to 2 decimal places, so only 1 mark awarded for method.

② Correct calculation per year, but needed to divide by 365 to calculate per day, and include units.

③ This is the correct equation but a comparison between the two terms should be included.

Ceri achieves 2/5 marks

Exam tip

It is important that you read the question carefully. Make sure that you follow all steps in any calculation.

Human impact on the environment

The following graph shows atmospheric carbon dioxide levels up to the year 2000.

Scientists have determined that the planetary boundary for climate change, which is determined by atmospheric carbon dioxide, is 350 ppm.

a) What is meant by a planetary boundary? Using information in the graph, explain why the climate change boundary has been crossed. (3)

b) Calculate the average rate of increase in atmospheric carbon dioxide between 1920 and 2000, and use this to estimate the atmospheric carbon dioxide concentration in 2030. (3)

c) Explain why the value you calculated in part b) is likely to be inaccurate, and how you could improve the accuracy of your calculation. (3)

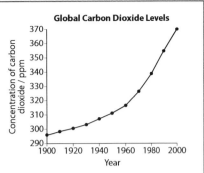

Global atmospheric carbon dioxide concentration

Lucie's answer

a) It is a framework proposed by environmental scientists that marks a safe zone for humanity as a precondition for sustainable development. ✓ The planetary boundary for climate change as represented by atmospheric carbon dioxide was crossed in the late 1980s. ✓ ①

b) 370 – 300 = 70 ppm ✓

$$\frac{70}{80} = 0.875 \text{ ppm yr}^{-1}.$$

0.875 × 30 = 26.3 ✓
26.3 + 370 = 396.3 ppm. ✓

c) The CO_2 concentration in 2015 already exceed this. ✓ I calculated the value using an average rate of increase between 1920 and 2000, but the rate of increase in CO_2 increased in the 1960s so my value will be lower. ✓ It would be better to calculate the rate from 1980–2000 and use this. ✓

Examiner commentary

① Lucie's answer could have been expanded to include the reasons, e.g. increased combustion of fossil fuels, deforestation, mechanised agriculture, use of fertilisers.

Lucie achieves 8/9 marks

Ceri's answer

a) They are proposed by scientists to mark a safe zone for the planet. ① The current atmospheric carbon dioxide is higher than 350 ppm. ✗ ②

b) 385 ppm. ✗ ③

c) The CO_2 concentration in 2015 is 400 ppm which already exceeds this. ✓ ④

Examiner commentary

① Ceri should include reference to sustainability.

② Information from the graph, e.g. that the value in 2000 was 370, or that 350 ppm was crossed in late 1980s should be included.

③ It looks like Ceri miscalculated the increase by incorrectly reading off the graph. Whilst Ceri has added 15 ppm to 370 ppm, without any working it is not possible to give any credit.

④ Ceri needs to explain where the inaccuracy may have come from, i.e. that the rate of increase changed during the period, or how it could be improved.

Ceri achieves 1/9 marks

Exam tip

Learn your definitions! It is important to show working in any calculations, so credit for method can be given. Identifying inaccuracies is a difficult skill, so look carefully for any anomalies and try to identify any assumptions that have been made.

Q&A 6

> **Essay 1**
> Explain the importance of different farming activities used to maximise efficient food production, and the consequences upon planetary boundaries of their use.

Lucie's answer

Farmers need to maximise plant growth in order to maximise food production. Plants need a source of nitrogen to synthesise proteins needed for growth and to make enzymes and hormones. Growth rates are increased in soils which have a good supply of nitrogen. ✓ Plants obtain nitrogen from the soil usually in the form of nitrates. In the soil, bacteria recycle nitrogen through the nitrogen cycle. Ploughing and drainage are both important because they aerate the soil. This is important because oxygen is needed for the active transport of mineral ions including nitrates into the plant roots. ✓ It also promotes nitrification where *Nitrosomonas* bacteria convert ammonium ions into nitrites, and *Nitrobacter* bacteria which convert nitrites into nitrates. ✓ Both of these bacteria respire aerobically and so need oxygen to do this. The process of denitrification which is carried out in the soil by *Pseudomonas* bacteria converting nitrates back into atmospheric nitrogen is an anaerobic process, and so is inhibited in well-aerated soils. Where soils lack nitrogen, farmers can plant leguminous plants such as peas and clover. ✓ These plants have root nodules which contain *Rhizobium* which are nitrogen fixing bacteria which are able to increase soil nitrogen content when they are ploughed back into the soil. ①
Farmers also can apply nitrogen-based fertilisers like ammonium nitrate, which is produced by the Haber process. ✓ This requires much energy from fossil fuels to make them which causes atmospheric pollution in the form of carbon dioxide, which is a greenhouse gas. Activities such as ploughing involve the use of machinery, which also causes carbon dioxide pollution. These activities have led to increases in carbon dioxide emissions, which has meant that the climate change boundary has been crossed. ✓ Farming and the removal of hedgerows to allow for ever larger machinery have led to the extinction of species as habitats such as hedgerows have been lost. This activity, and other habitat losses have led to the biodiversity boundary being exceeded. ✓ The excessive removal of atmospheric nitrogen during the Haber process has resulted in the biochemical boundary for nitrogen being crossed. ✓ ②

Examiner commentary

Lucie gives a full and detailed account of the different farming activities and how they influence the nitrogen cycle. The effect on three planetary boundaries has been discussed. The account is articulate and shows sequential reasoning. There are no significant omissions.

① Lucie could have included the use of manure and its breakdown through ammonification to increase nitrogen content.

② Lucie could have included a definition of what is meant by a planetary boundary. This then would put into context why exceeding the boundaries is an important consequence.

Lucie achieves 8/9.

Ceri's answer

Plants need nitrogen to grow. It is needed to manufacture proteins. ① Plants take up nitrogen in the form of nitrates from the soil. Farmers can do a lot to help increase the amount of nitrogen in the soil in order to maximise growth of crops, for example they can add fertilisers and manure. ✓ There are bacteria in the soil that help to break down organic waste into nitrates by a process called nitrification, e.g. *Nitrosomonas* and *Nitrobacter*. ② Ploughing also helps as it mixes manure through the soil, improving soil texture and improves oxygenation of soil ③ A soil rich in oxygen also inhibits denitrification where nitrates are converted back into atmospheric nitrogen by bacteria called *Pseudomonas denitrificans*. ✓ The overuse of inorganic fertilisers has affected a number of planetary boundaries, e.g. Biochemical boundary for nitrogen which has been exceeded. ✓ ④ Monoculture has also affected the biodiversity boundary. ⑤

Examiner commentary

Ceri gives a limited account of the different farming activities and how they influence soil fertility. Two planetary boundaries have been discussed but the effect on the biochemical boundary is not mentioned. Ceri makes some relevant points, correctly names three bacterial species involved in the nitrogen cycle but could have detailed nitrification more clearly. There is limited use of scientific vocabulary.

① Ceri needs to explain in more detail why proteins are needed, e.g. to synthesise enzymes.

② This lacks detail. The answer does not explain that *Nitrosomonas* bacteria convert ammonium ions into nitrites, and *Nitrobacter* bacteria which convert nitrites into nitrates.

③ There is no mention of drainage, or why increased oxygenation improves soil nitrogen, i.e. that nitrification is an aerobic process and that nitrate uptake by roots requires oxygen for active transport. The use of leguminous plants and nitrogen fixation is omitted.

④ There is no mention of climate change boundary being exceeded or why.

⑤ There is no detail on why and how the biodiversity boundary has been exceeded, e.g. due to habitat loss from hedgerow removal.

Ceri achieves 3/9

> **Exam tip**
>
> Remember it is not a mark per point, but rather what you say and how you say it. Answers must not omit any key information, and should include all key scientific terminology to gain top marks. Watch your spelling too!

Homeostasis and the kidney

The table below shows the typical concentrations of two solutes (glucose and urea) in three different regions of the kidney nephron, labelled P, R and S, in the diagram below.

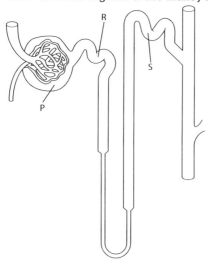

Solute	Mean concentration of solute / g dm^{-3}		
	P	**R**	**S**
Glucose	0.12	0.00	0.00
Urea	0.35	0.65	6.25

a) State exactly where you would expect to find the structure labelled P in a cross section through the kidney. (1)

b) Explain how the changes in concentration for glucose and urea are brought about. (6)

Lucie's answer

a) In the cortex. ✓

b) The mean concentration of glucose decreases from 0.12 g dm^{-3} in region P to 0.00 g dm^{-3} in regions R and S ✓ because glucose is selectively reabsorbed into the blood in region R. ✓ Urea is not selectively reabsorbed in this region, but water is, ✓ so the concentration of urea increases from 0.35 g dm^{-3} to 6.25 g dm^{-3} because the same mass of urea is dissolved in a smaller volume of water. ✓ ①

Examiner commentary

① Lucie should have included details of the mechanism of reabsorption here, i.e. glucose is reabsorbed by co-transport with sodium ions and water by osmosis.

Lucie achieves 5/7

Ceri's answer

a) Cortex. ✓

b) Glucose is reabsorbed by co-transport with sodium ions ✓ causing its concentration to decrease. ✓ ① Urea is not reabsorbed in regions R and S ✓ which is why its concentration increases. ✗ ②

Examiner commentary

① It is important to quote data when explaining.

② The reason given is incorrect: Water is selectively reabsorbed by osmosis in the proximal convoluted tubule and the loop of Henle, so the same mass of urea is dissolved in a smaller volume of water causing its concentration to increase.

Ceri achieves 4/7

Exam tip

Explain answers fully, and always quote data to support your answer.

The nervous system

The diagram shows a typical reflex arc found in the mammalian nervous system.

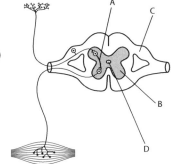

a) Complete the table, naming the structures A–D. (3)

Letter	Name
A	
B	
C	
D	

b) The diagram below shows a section through a motor neurone.

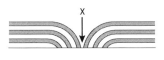

Identify X and Y and explain how they affect the speed of nerve impulse transmission. (4)

Lucie's answer

a) A is relay neurone, ✓ B is Grey matter, ✓ C is dorsal root ✗ ① D is central canal. ✓ (max 2 out of 3 allowed)

b) X is the node of Ranvier, and Y is the myelin sheath ✓. The presence of the myelin sheath speeds up impulse transmission because ions can only pass across the membrane at the nodes of Ranvier where there is no myelin, ✓ so depolarisation only occurs at these nodes. ✓ The impulse jumps from node to node. ✗ ②

Examiner commentary

① C is pointing to the swelling in the dorsal root which is called the dorsal root ganglion.

② The action potential 'jumps' from node to node, **not** the impulse.

Lucie achieves 5/7

Ceri's answer

a) A = relay neurone, ✓ B= grey matter, ✓ C= ganglion✗ ① D= cerebro-spinal fluid✗ ② (max 1 out of 3)

b) The myelin sheath (Y) ③ speeds up impulse transmission because depolarisation can only occur at these gaps in the myelin sheath✓ so the action potential jumps from node to node. ✓④

Examiner commentary

① C= dorsal root ganglion

② The central canal contains cerebrospinal fluid.

③ X is not named as node of Ranvier (both are required for one mark).

④ Ceri should have included reference to ion movement only being possible at the nodes or that local circuits set up over a larger distance.

Ceri achieves 3/7

Exam tip

Label your diagrams correctly! Read the pointers in the guide: They point out common mistakes to avoid like impulse 'jumps'.

Q&A 9

Essay 2

Parkinson's disease is a progressive neurological condition resulting from the death of brain cells that produce dopamine, the neurotransmitter involved in motor control pathways in the brain. Patients are unable to control fine motor movements such as walking. Treatment involves the use of L-dopa which is a synthetic drug that is converted into dopamine in the brain. The decarboxylation of L-dopa into dopamine is shown below.

HO, NH$_2$, CO$_2$H → HO, NH$_2$

Using the information, describe synaptic transmission and explain the use of Levodopa (L-dopa) to treat sufferers of Parkinson's disease.   (9 marks)

Lucie's answer

When an action potential arrives at a pre-synaptic neurone, calcium channels in the membrane open, causing calcium ions to rush into the synaptic knob. ✓This causes synaptic vesicles to migrate to the pre-synaptic membrane and to fuse with it. ✓The neurotransmitter is released into the synaptic cleft by exocytosis. ✓The neurotransmitter diffuses across the cleft and binds to receptors on sodium channels located on the post-synaptic membrane causing them to open. ✓Sodium ions rush into the post-synaptic neurone depolarising it. This triggers an action potential in the post-synaptic neurone. ✓ In Parkinson's disease, the death of brain cells results in less dopamine being released. Dopamine is a neurotransmitter involved in synapses in the brain responsible for fine motor control. Because less neurotransmitter is released, fewer post-synaptic neurones will be depolarised which will cause fewer muscles fibres to contract resulting in difficulty in walking. ✓ L-dopa is decarboxylated ✓ into dopamine by a one step reaction in the brain, supplying dopamine quickly to the affected areas. The increase in neurotransmitter allows more post-synaptic neurones to be depolarised, so more muscles fibres contract, making walking easier. ✓

Examiner commentary

Lucie gives a full description of synaptic transmission, and explains the use of L-dopa well. The account is articulate and shows sequential reasoning. There are no significant omissions but Lucie should make it clear that brain synapses are being discussed.

Lucie achieves 8/9

Ceri's answer

Calcium channels in the membrane open, so calcium ions diffuse ① into the synaptic knob which triggers release of neurotransmitter. ② Neurotransmitter molecules diffuse across the cleft and bind to receptors on ③ the post-synaptic membrane causing sodium channels to open. ✓ Sodium ions diffuse into the post-synaptic neurone which triggers an action potential. ④ People who suffer from Parkinson's disease, don't produce enough dopamine because brain cells that produce it have died. This results in poor motor control because fewer motor neurones will be depolarised. ✓ ⑤ L-dopa works as a precursor to dopamine and is easily converted into dopamine in the brain. ⑥ The increased levels of dopamine restore function allowing more motor neurones to be depolarised which makes walking easier. ✓

Examiner commentary

Ceri gives a sound description of synaptic transmission but some steps are not fully detailed. Ceri makes some relevant points, but there is limited use of scientific vocabulary.

① Movement of ions is rapid so Ceri should indicate that ions rush in or diffuse in rapidly.

② Ceri should include that vesicles containing neurotransmitter migrate to and fuse with the pre-synaptic membrane.

③ Receptors on the sodium channels.

④ It is better to say depolarise the post-synaptic membrane which establishes an action potential.

⑤ Need to link to muscle contractions – that fewer muscle fibres will contract.

⑥ Ceri should use information in the diagram, i.e. the loss of carbon dioxide which is decarboxylation.

Ceri achieves 3/9

> **Exam tip**
> Review all the information provided in the question and decide which information is relevant to the part you are answering.

Unit 4

Q&A 10

Sexual reproduction

a) Identify cells B and C, and explain how cell C differs from cell B. (3)

b) Identify cell D and describe the process that happens to cell D to enable it to fertilise a secondary oocyte. (4)

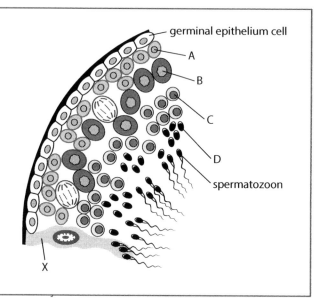

germinal epithelium cell
A
B
C
D
spermatozoon
X

Lucie's answer

a) Cell C is a secondary spermatocyte and is haploid, ✓ whereas cell B is a primary spermatocyte and is diploid ✓ and undergoes meiosis I to produce cell C. ✓

b) Cell D which is a spermatid undergoes differentiation. ✓ This involves no change to the chromosomes in the cell but incorporation of an acrosome in the head of the spermatozoa, containing hydrolytic enzymes which allows it to digest the zone pellucida of the ovum. ✓ ① A mid piece is added containing numerous mitochondria ② and a tail which provides movement toward the secondary oocyte. ✓

Examiner commentary

① Prior to the sperm entering zona pellucida, the female gamete is referred to as the secondary oocyte as meiosis II hasn't occurred yet.

② Lucie should include the role of the numerous mitochondria, i.e. to provide the ATP necessary for locomotion.

Lucie achieves 6/7

Ceri's answer

a) Cell C is a secondary spermatocyte and undergoes meiosis I ✓ to produce cell B which is a primary spermatocyte. ✓ ①

b) Cell D which is a spermatocyte ✗ ② and undergoes differentiation. The spermatozoa has a head containing an acrosome, ③ a mid piece containing mitochondria ③ and a tail for movement. ✓

Examiner commentary

① Ceri needs to include the main difference between the two cells, i.e. that C is diploid and that B is haploid.

② Cell D is a spermatid.

③ Ceri needs to explain the function of these parts of the spermatozoa enabling fertilisation of the secondary oocyte, i.e. acrosome contains hydrolytic enzymes which digest the zona pellucida and mitochondria that produce ATP for movement.

Ceri achieves 3/7

Exam tip

Link structure to function, i.e. explain the function.

Inheritance

In Guinea pigs the allele for black coat (B) is dominant to the allele for albino (b) and the allele for rough coat (R) is dominant to the allele for smooth coat (r). A heterozygous black smooth coated guinea pig was mated with an albino guinea pig which is heterozygous for rough coat. In the first generation the offspring (F$_1$) had the following phenotypes: 27 black rough coat; 22 black smooth coat; 28 albino rough coat; 23 albino smooth coat.

a) Complete the genetic diagram below to show how the offspring in the first generation inherited the phenotype a shown above.   (5)

Parental genotype.   X   ....

Gametes   X   ....

F$_1$ genotypes   ....

F$_2$ phenotypes   ....

Phenotype ratio   ....

b) Using the table below calculate χ^2 for the results of the cross.   (3)

Category	Observed O	Expected E			
Black rough coat					
Black smooth coat					
Albino rough coat					
Albino smooth coat					

$$\chi^2 = \sum \frac{(0 - E)^2}{E}$$

$\chi^2 = $ ....

c) Use your calculated χ^2 value and the probability table to conclude how coat color and texture are inherited.   (4)

Degrees of freedom	p= 0.10	p= 0.05	p= 0.02
1	2.71	3.84	5.41
2	4.61	5.99	7.82
3	6.25	7.82	9.84
4	7.78	9.49	11.67
5	9.24	11.07	13.39

Lucie's answer

a) Bbrr   X   bbRr   ✓
   Br, br   bR, br   ✓

	Br	br	
bR	BbRr	bbRr	✓
	black rough	albino rough	
br	Bbrr	bbrr	✓
	black smooth	albino smooth	

Ratio is 1:1:1:1 ✓

b)

Category	Observed O	Expected E	O–E	$(O–E)^2$	$(O–E)^2 / E$
Black rough coat	27	25	2	4	0.16
Black smooth coat	22	25	–3	9	0.36
Albino rough coat	28	25	3	9	0.36
Albino smooth coat	23	25	–2	4	0.16
Σ	100	100			1.04

                           ✓   ✓

$\chi^2 = 1.04$ ✓

C) The null hypothesis is that there is no significant difference between the observed and expected values. ✓

Because the calculated value 1.04 is less than the critical value at p=0.05, 7.82, we can accept the null hypothesis, so any differences between observed and expected results seen were due to chance. ✓

①

Examiner commentary

① Lucie should include how coat colour and texture are inherited, i.e. Mendelian genetics therefore applies and guinea pig coat colour and texture genes are not linked. Coat colour is controlled by a dominant black allele and a recessive albino allele, coat texture is controlled by a dominant rough allele and a recessive smooth allele.

Lucie achieves 10/12

Ceri's answer

a) Bbrr   X   bbRr   ✓
    Br, br   bR, br   ✓

	Br	br	
bR	BbRr	bbRr	✓
br	Bbrr	bbrr	①

Ratio is 1:1:1:1 ✓

b)

Category	Observed O	Expected E	O–E	(O–E)²	
Black rough coat	27	25	2	4	
Black smooth coat	22	25	–3	9	
Albino rough coat	28	25	3	9	
Albino smooth coat	23	25	–2	4	
$\sum$	100	100		26	

✓

$$\frac{26}{100} = 0.26$$

$\chi^2 = 0.26$ ✗ ②

c) Because the calculated value 0.26 is less than the critical value of 7.82, ③ we can accept the null hypothesis, so any differences seen were due to chance. ✓
    ④

Examiner commentary

① Ceri should include phenotypes either in the table or in the ratios.

② Ceri has summed (O–E)² and then divided this by the sum of E rather that working out each (O–E)² / E and summing these. The calculated chi squared value is wrong as a result.

③ Ceri must include the level of probability used, i.e. p=0.05, as 7.82 is also the value for 2 degrees of freedom at p = 0.02.

④ Ceri has been awarded error carried forward – even though the calculated value is wrong, this was penalised in part b) so is not penalised again in part c). Ceri needs to include a null hypothesis, and should include how coat colour and texture are inherited, i.e. Mendelian genetics therefore applies and guinea pig coat colour and texture genes are not linked, etc.

Ceri achieves 6/12

> **Exam tip**
> When performing statistical tests, include a null hypothesis and ensure that results are explained in terms of significance, chance and the probability value 0.05.

Q & A 12

Variation and evolution

Thalassemia is the name of a group of inherited conditions that affects the production of haemoglobin. It is caused by a recessive allele which results in too little haemoglobin being produced which causes anaemia, and shortness of breath. It is most common in people with Mediterranean or Asian origins, affecting 1 in 2000 babies screened In the UK.

The Hardy–Weinberg formula states that if alleles A and a are present in a population with the frequencies of p and q, the proportion of individuals homozygous for the dominant allele (AA) will be p^2, the proportion of heterozygotes (Aa) will be 2pq, and the proportion or homozygous recessives (aa) will be q^2, where $p + q = 1$.

a)   What is meant by the term recessive allele?   (2)

b)   Use the Hardy–Weinberg formula to estimate the number of carriers of thalassemia per 1000 in the UK. Show your working.   (4)

Lucie's answer

a) An allele that is only expressed in the homozygous recessive, e.g. aa. ✓ ①

b) $aa = \dfrac{1}{2000} = 0.0005 = q^2$ ✓

  $q = \sqrt{0.0005} = 0.022$

  $p = 1 - 0.022 = 0.978$. ✓

  $Aa = 2pq = 2 \times 0.022 \times 0.978$ ✓ $= 0.043$ or 43 per 1000 population. ✓

Ceri's answer

a) A recessive allele codes for a protein ✗ ① and the phenotype is only seen when both copies of the allele are present, e.g. aa. ✓

b) $q^2 = \dfrac{1}{2000} = 0.0005$ ✓

  $q = 0.022$

  $p = 0.978$. ✓

  proportion of population that are carriers are 0.978 ②

Examiner commentary

① Lucie needs to define allele, i.e. a different form of the same gene (a gene is a section of DNA that codes for a specific polypeptide).

Lucie achieves 5/6

Examiner commentary

① Ceri needs to define allele – that it is a different form of the same gene.

② Ceri should show workings more fully to enable marks for process to be given. The final step of multiplying 0.978 by 1000 to get the proportion of the population per 1000 needs to be included.

Ceri achieves 3/6

Q&A 13

Application of reproduction and genetics

Scientists started mapping a section of DNA by digesting it with different restriction enzymes and estimating the size of each fragment by running the products out on an agarose gel alongside a DNA ladder containing DNA fragments of known sizes. The results are shown in the table below.

Enzymes used	Estimated sizes of fragments produced /base pairs
EcoRI	550, 450
BamHI	750, 300
SnaI	500, 325, 200
EcoRI and PstII	550, 450
EcoRI and HindIII	550, 250, 200

a) What is a restriction enzyme? (1)
b) Using a DNA ladder to estimate size of DNA fragments has its limitations and is often inaccurate. What evidence is there in the data to support this claim? (2)
c) Draw conclusions from the results, justifying your answer. (3)

Lucie's answer

a) A bacterial enzyme that cuts single stranded DNA at a specific base pair sequence. ✓

b) The same DNA was cut with different enzymes and the total size of the fragments produced by each reaction was different ✓, e.g. EcoRI produced fragments totaling 1000, whereas BamHI totalled 1050 even though the DNA used was the same. ✓

c) The size of the DNA fragment is 1000 because the fragments produced in all the digests total approximately 1000. ✓ PstII does not cut the DNA so there is no recognition sequence for PstII in the sample, because the number and size of fragments produced are the same as when using EcoRI alone. ✓ HindII cuts within the 450 bp EcoRI fragment because when both enzymes are used the 450 bp fragment is no longer present, but two fragments totaling 450 bp are present. ✓ ①

Ceri's answer

a) An enzyme that cuts DNA. ①

b) Different sized fragments are produced when different enzymes are used. ②

c) The DNA must be approximately 1000 bases long. ③ There is no site for PstII to cut within the DNA, because the result is the same when using EcoRI and EcoRI & PstII. ✓ The other enzymes cut once but SnaI must cut twice because three fragments are produced. ✓

Examiner commentary

① Lucie could also conclude that EcoRI and BamHI only cut the DNA once, as two fragments are produced. SnaI cuts twice as three fragments are produced.

Lucie achieves 6/6

Examiner commentary

① The definition needs further detail, e.g. Cuts DNA at a specific recognition sequence.

② Ceri needs to include specific examples to support the answer, e.g. fragments total 1000 base pairs when cut with EcoRI but 1025 when cut with SnaI.

③ The conclusions are valid but should include justification, e.g. DNA is approx 1000 bases long because the fragments produced add to 1000 with EcoRI but 1025 with SnaI.

Ceri achieves 2/6

Exam tip

Always use evidence to justify conclusions.

Q & A 14

Essay 3

An experiment was carried out to investigate the optimum requirements for germinating broad bean seeds. Ten seeds were placed in the different environments shown in the table below and the mean seedling height was recorded ten days after germination.

Temperature / °C	Volume of water supplied per day/cm³	Mean seedling height / mm
10	15	41
10	30	48
10	60	9
20	15	65
20	30	72
20	60	13
30	15	82
30	30	86
30	60	18

Using the results, draw a conclusion as to the optimum conditions required for seedling growth in the broad bean. Using your biological knowledge, explain your conclusion.

(9)

Lucie's answer

The results show that both temperature and water affect growth in germinating broad bean seedlings. The optimum conditions for growth are 30°C and 30 cm³ of water supplied, which produced a mean seedling height of 86mm after ten days. ✓ Too much water inhibited growth, which was most noticeable at 10°C where the mean seedling height for 60 cm³ was just 9 mm. ✓ ①

Water is needed for germination and seedling growth. Water is absorbed by the seed causing the testa to split as the tissues swell. The radicle emerges and begins to absorb more water. Water is important because it mobilises enzymes and provides water for the hydrolysis of starch into maltose. ✓ ② Once the seedling is photosynthesising, water is also needed for photosynthesis, and transport of sucrose to the growing points. ✓ An optimum temperature is important not only for enzymes because temperature increases the kinetic energy of enzyme and substrate molecules resulting in more enzyme–substrate complexes, but for the enzymes involved in photosynthesis, e.g. RuBisCo. ✓ Too much water inhibited growth, for example even at the optimum of 30°C, increasing the water supplied from 30 to 60 cm³ caused a reduction in height of seedling from 86 to 18 mm. ✓ Water is needed for growth, but too much reduces the oxygen available to the roots of the developing seedling, as water replaces air in the air spaces found within the soil. Oxygen is needed for aerobic respiration of maltose within the germinating seed and later for the active uptake of mineral ions, e.g. nitrates into the roots. Nitrates are essential for growth because they are needed by plants to synthesise proteins, e.g. enzymes and structural proteins. ✓

Examiner commentary

Lucie gives a full conclusion which is supported by detailed biological knowledge from both AS and A2. The account is articulate and shows sequential reasoning. There are no significant omissions.

① Data is quoted, but some could have been processed, e.g. % increase. ② hydrolysis of maltose is mentioned, but Lucie could have included an equation or talked more about the chemical addition of water breaking the glycosidic bonds. It would be better to say that hydrolysis of maltose is followed by respiration of glucose. Reference to oxygen being needed for aerobic respiration to fuel biosynthesis for growth could be included.

Lucie achieves 7/9

Ceri's answer

The best conditions for growth were 30°C and 30 cm³ of water supplied. ✓ ①

Water is important because it is needed to hydrolyse starch into maltose within the seed and is needed for photosynthesis. ② 30°C provides the best growth because temperature increases the kinetic energy of enzyme and substrate molecules resulting in more enzyme-substrate complexes. ✓ ③ Above 30 cm³ height of seedling was reduced, perhaps because too much water reduces the oxygen in the soil. ④ Oxygen is needed for aerobic respiration in the seed. Aerobic respiration yields more ATP than anaerobic respiration. ⑤

Examiner commentary

Ceri gives a brief conclusion and explanation of the results. Ceri makes some relevant points, but there is limited use of scientific vocabulary.

① Data must be quoted, e.g. 86 mm height, and the conclusion needs to be more detailed.

② Ceri should include details of hydrolysis of maltose and why water is needed for photosynthesis, and translocation of solutes, etc.

③ Ceri should name some enzymes, e.g. maltase, RuBisCO.

④ Ceri should include why oxygen is important to the growing seedling, e.g. active uptake of nitrates and other mineral ions.

⑤ Ceri should develop this further that less ATP would result in less uptake of nitrates or protein synthesis in the seedling.

Ceri achieves 2/9

Exam tip

Be specific and name both enzyme and substrate. Always include data to support your conclusion.

Option A

Q&A 15

Immunity and Disease

The graph shows the antibody concentration in the blood following two exposures to the same antigen.

a) Explain why the antibody concentration in the blood is higher following the second exposure than the first exposure. (3)

b) Using examples, explain two differences between active and passive immunity. (2)

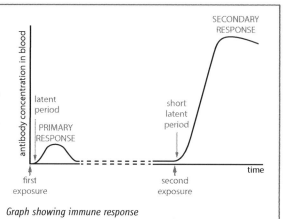

Graph showing immune response

Lucie's answer

a) Following first exposure, macrophages need to engulf the foreign antigen and incorporate the antigens into their own cell membranes during the time called the latent period. ✓ ① Following second exposure, memory cells undergo clonal expansion much faster than following the first exposure because antigen presentation doesn't occur, so antibodies are made much more quickly and in larger quantities. ✓

b) In active immunity a person produces antibodies in response to infection, e.g. measles or vaccination, e.g. MMR, whereas in passive immunity a person receives antibodies either from the breast milk or from an antibody injection, e.g. Rabies vaccination. ✓ Active immunity gives longer lasting protection than passive because memory cells are produced. ✓

Examiner commentary

① Lucie should include that T helper cells secrete cytokines which trigger B plasma cells to produce antibodies, which takes time.

Lucie achieves 4/5

Ceri's answer

a) Concentration is higher following second exposure because memory cells undergo clonal expansion much faster than following the first exposure because macrophages don't need to engulf the foreign antigen and incorporate the antigens into their own cell membranes. ✓ ①

b) Active immunity involves an infection whereas passive immunity involves an injection. ✗ ② Active immunity gives longer lasting protection than passive immunity. ③

Examiner commentary

① Ceri needs to include that during the primary response T helper cells need to secrete cytokines to trigger B plasma cells to produce antibodies which takes time. Because the secondary response occurs more quickly, a higher level of antibody production can be achieved.

② Ceri needs to be clear about an injection of antibodies and a vaccination which is an injection of antigens and include examples, e.g. MMR vaccine versus an injection of antibodies, e.g. to treat rabies.

③ Ceri needs to include a reason, e.g. active immunity gives longer lasting protection because memory cells are produced.

Ceri achieves 1/5

Exam tip
It is important to make a comparison when differences are asked for, and to include examples.

Option B

Q&A 16

Human Musculoskeletal Anatomy

The following diagram shows a section through skeletal muscle.

a) Label the diagram to clearly show:
   i)   M line
   ii) Z line
   iii) I band
   iv) **One** sarcomere   (3)

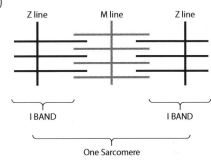

Section through skeletal muscle

b) Describe what would happen to the section of muscle shown above following contraction.   (2)

c) During strenuous exercise muscles may temporarily respire anaerobically. Explain why it is important for the muscles of an athlete to convert pyruvate into lactate (lactic acid) and the consequence of a build-up of lactate on muscle contraction.   (3)

Lucie's answer

a)

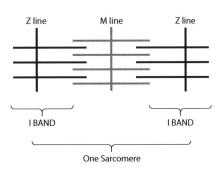

✓✓✓

b) Sarcomere shortens as I band and H zone shorten. ✓ A band remains the same length. ✓

c) it allows allowing glycolysis to continue, because NAD is regenerated ✓ when pyruvate is reduced to lactate. ✓ ①

Examiner commentary

① Lucie needs to include the effect of a build up of lactate upon muscle contraction, i.e. that it inhibits chloride ions which regulate muscle contraction resulting in a sustained contraction leading to cramp.

Lucie achieves 7/8

Ceri's answer

a)

Z line   M line   Z line

I BAND   I BAND

One Sarcomere

✓✓✓

b) I band shortens as does the sarcomere. ✓ ①

c) It allows for the build up of an oxygen debt. ②

Examiner commentary

① Ceri needs to include the effect on the A band, i.e. that the A band remains the same length.

② Whilst a build-up of an oxygen debt happens Ceri needs to go further to explain the consequence, i.e. that NAD is regenerated allowing glycolysis to continue. Ceri also needs to include the consequence of a build-up of lactate on muscle contraction.

Ceri achieves 4/8

Exam tip

Read the question carefully and explain your answer fully.

Option C

Neurobiology and Behaviour

a) Explain the difference between taxes and kineses. (1)

b) A victim of a car crash suffered damage to their forebrain. Explain why this person would have difficulties in forming permanent memories, but would be less likely to suffer from stress. (2)

c) Use examples to distinguish between classical and operant conditioning. (2)

Lucie's answer

a) Kinesis are non-directional whereas in taxes the direction of the movement is related to the direction of the stimulus either towards or away from it. ✓

b) The hippocampus is located in the forebrain and is involved in consolidating memories into a permanent store. If this was damaged the person would be unable to do this. ✓ It is also responsible for producing cortisol, the stress hormone. ①

c) Classical conditioning involves the association between a natural and an artificial stimulus to bring about the same response, e.g. a dog associating the ringing of a bell with food, ✓ whereas operant conditioning involves the association between a particular behaviour and a reward or punishment, e.g. mice learning to press a lever to receive food (reward) or to stop a loud noise (punishment). ✓

Ceri's answer

a) Kinesis are non-directional; taxes are directional. ①

b) The hippocampus may have been damaged, which is responsible for forming memories ✓ It is also responsible for producing cortisol. ②

c) In classical conditioning a dog learns to associate the ringing of a bell with food, whereas operant conditioning involves mice learning to press a lever to receive food. ✓ ③

Examiner commentary

① Lucie needs to make a clear link between damage to the hippocampus and its role in controlling the production of cortisol from the adrenal glands.

Lucie achieves 4/5

Examiner commentary

① Taxes involve movement that is related to the direction of the stimulus.

② The role of cortisol needs to be included.

③ Ceri should include the main difference between the two types of conditioning, i.e. that classical conditioning involves the association between a natural and an artificial stimulus to bring about the same response whereas operant conditioning involves the association between a particular behaviour and a reward or punishment.

Ceri achieves 2/5

Exam tip

When distinguishing between two terms make sure you include detail on both terms.

Additional practice questions

1. a) Name the group of biological molecules to which ATP belongs. (1)

   b) Explain the role of water in the light-dependent stage of photosynthesis. (3)

2. a) Name the processes by which:

   i) Ammonia is converted into nitrates by bacteria in the soil. (1)

   ii) Nitrates are converted into atmospheric nitrogen. (1)

   b) Describe how atmospheric nitrogen can be converted directly into nitrogen compounds for plants. (3)

3. a) Describe two differences between slow twitch and fast twitch muscle fibres. (2)

   b) The type of training used by marathon runners has been shown to increase the relative proportion of slow twitch fibres. State one other change that occurs in muscles during endurance training, and explain the benefit to a marathon runner. (3)

4. A student used the Gram stain to stain the bacteria to make them more visible under a light microscope. The bacteria were all spherical in shape, but some appeared purple, others pink.

   a) Identify the type of bacteria which retained the purple stain. (2)

   b) Explain why some bacteria stained purple whilst others stained pink. (3)

5. Aerobic respiration occurs in a number of stages.

   a) Complete the table using a tick (✓) to indicate which statements apply to the following stages in respiration, or a cross (✗) if they do not. (4)

Statement	Glycolysis	Link reaction	Krebs cycle	Electron transport chain
Occurs in the mitochondrial matrix				
ATP produced by substrate level phosphorylation				
FAD reduced				
Reduced NAD is oxidised				

   b) Explain the role of ATP in glycolysis. (2)

6. a) Describe what is meant by resting potential. (2)

   b) Explain how the resting potential is maintained across a neurone. (3)

7. Explain how cells lining the proximal convoluted tubule are adapted for selective reabsorption. (3)

8. a) State one similarity between the process where the sperm nucleus enters the secondary oocyte in humans and the male nucleus enters a plant ovule. (1)

   b) Outline the main differences between meiosis in spermatogenesis and oogenesis. (3)

9. a) Distinguish between gene and allele. (1)

   b) A pea plant produced yellow wrinkled peas. A scientist predicted using Mendel's second law of inheritance that the plant was heterozygous for colour and texture, but when he self-pollinated the plant, 96 of the offspring were yellow wrinkled and 32 were green smooth.
      i) State Mendel's second law of inheritance. (1)
      ii) Explain what could have happened to produce the results seen. (4)

10. a) Distinguish between continuous and discontinuous variation. (3)

    b) Fur colour in a species of mouse is controlled by one gene with two alleles. Scientists used the Hardy–Weinberg equation to estimate the allele frequency of the dominant allele that produces dark fur, and the recessive allele that produces light fur colour in different environments where the mouse was found. In rocky environments, the frequency of the dark allele was much higher than in sandy environments where the light allele frequency was higher, and so they concluded that this had arisen through natural selection. Explain how different fur colour could have arisen in the two different environments. (4)

11. a) Explain how gel electrophoresis can be used to separate DNA fragments according to their size. (3)

    b) Using your knowledge of sickle cell disease, and gene technology, briefly outline how you *might in theory* treat a sufferer of sickle cell using gene therapy. (3)

12. Explain the treatments available for kidney failure. (9) QER

13. Antibodies are produced in response to foreign antigens and are Y-shaped. The amino acid sequence in the variable region varies greatly between different antibodies. This variable region, composed of 110–130 amino acids, gives the antibody its specificity for binding to an antigen. Vaccinations are widely used to give protection against a range of diseases by triggering antibody production in the recipient. Antibody responses to influenza vaccine in HIV-infected persons tend to be impaired, with more severe impairment in later stages of HIV infection.

    a) Name the immunological response that produces antibodies. (1)

    b) A patient with AIDS would be unable to produce sufficient antibodies to give protection against influenza. Explain why. (3)

14. a) Describe the functions of the hindbrain. (2)

    b) Outline the techniques that could be used to identify *functional* brain damage following a stroke. (6)

15. Demyelination is the loss of the myelin sheath, which occurs in some neurodegenerative autoimmune diseases, including multiple sclerosis. This condition damages the myelin sheath of neurones in the brain, spinal cord and optic nerves. The reaction times of multiple sclerosis sufferers were compared with healthy individuals, together with the percentage of correct responses to a visual stimulus. The results are shown below.

Group	Reaction time /ms	% of correct responses
Multiple sclerosis sufferer	549	80
Control group (healthy)	443	95

    i) Calculate the percentage increase in reaction time for the group with multiple sclerosis compared to a healthy individual. Show your working. (2)

    ii) Explain the difference in the reaction time results seen. (4)

    iii) Suggest a reason for decreased correct responses in multiple sclerosis sufferers. (1)

Answers to Quickfire and Extra questions

Unit 3

① Movement in osmosis is derived from a water potential gradient whereas in chemiosmosis it is derived from a proton gradient.

②

Feature	Mitochondria	Chloroplasts
Site of electron transport chain	Cristae	
Co-enzyme involved		NADP
Terminal electron acceptor	Oxygen and H^+	

③ Any metabolically active cell, e.g. muscle or liver cell.

④ NADPH/reduced NADP and ATP.

⑤ Chloroplasts have a large surface area / they are able to move within palisade cells / contain several pigments / ANY 2.

⑥ It is able to convert light energy into chemical energy.

⑦ Chlorophyll b.

⑧ The absorption spectrum shows how much light energy is absorbed at different wavelengths whereas the action spectrum shows the rate of photosynthesis at different wavelengths.

⑨ Chlorophyll a.

⑩ Collect light energy at different wavelengths and funnel it to the reaction centre.

⑪ Thylakoid membranes in the chloroplast.

⑫ Any three: Cyclic involves just photosystem I whereas non-cyclic involves both photosystems. Cyclic produces 1 ATP, whereas non-cyclic produces 2 ATP and reduced NADP (NADPH). Electrons take a cyclical pathway in cyclic, but a linear one in non-cyclic. Cyclic does not generate oxygen as photolysis does not occur, unlike non-cyclic which liberates oxygen as a result of photolysis.

⑬ From the photolysis of water.

⑭ ATP and NADPH are needed, which are made by the light-dependent reaction. Without light, the stomata will close, limiting the uptake of carbon dioxide.

⑮ Using hydrogen from NADPH.

⑯ RuBisCO.

⑰ It is the 5 carbon acceptor molecule that fixes carbon dioxide. Without it the cycle could not continue.

⑱ $3 \times 2 = 6$ ATP from NADH plus net gain of 2 ATP produced directly = 8 ATP.

⑲ Increases the surface area for the attachment of enzymes, e.g. ATP synthetase.

⑳ Oxygen acts as the terminal electron acceptor. Without it electrons cannot leave the last proton pump, so no electrochemical/proton gradient is created.

㉑ a) cell cytoplasm, b) mitochondrial matrix, c) cell cytoplasm

㉒ Efficiency of anaerobic respiration

$$= \frac{30.6 \times 2}{2880} \times 100 = 2.1 \text{ % (1 dp)}$$

㉓ 1g of fat, because it contains large numbers of hydrogen atoms that are used to drive oxidative phosphorylation.

㉔ Spherical.

㉕ Stain purple. This is because they have a thick layer of peptidoglycan, which retains the crystal violet stain.

㉖ Lipopolysaccharide.

㉗ Nitrogen is needed for synthesis of nucleotides and proteins.

㉘ Total cell count include living and dead cells whereas viable cell count includes only living cells.

㉙ a) That every colony grew from a single bacterium.
b) $20 \times 1000 = 20\,000$ per 0.1 cm^3
$= 20\,000 \times 10 = 200\,000$ per cm^3 original culture
$= 200\,000 \times 25 = 5\,000\,000$ or
5×10^6 (1 dp).

㉚

Phase	Bacterial population	Mammalian population
Lag	Cells are growing, but little increase in number as time is needed for enzyme synthesis.	Slow growth represents the time taken to reach sexual maturity and to gestate young.
Log	Plenty of nutrients and few toxic by-products so there are no limiting factors.	Numbers increase logarithmically as there are no factors limiting growth.
Stationary	Cells are reproducing but population is constant due to cell production equalling cell death. The population has reached its carrying capacity due to reduced resources, e.g. nutrients/ space/ toxic waste products are now limiting factors.	Birth and death rates are equal and the population has reached its maximum size, or carrying capacity, due to reduced resources, e.g. food availability, or space, disease and predation.
Death	More cells are dying than are being produced so the population decreases. Death of cells is due to lack of nutrients, lack of O_2 or increased toxicity of the medium.	Death rate exceeds birth rate due to food availability, space, disease and predation.

㉛ $= \dfrac{56 \times 44}{22} = 112$

㉜ No births/deaths/ immigration/ emigration, have occurred during the time between collecting both samples. Behaviour not altered / animal not harmed by marking and released animals fully integrated back into population.

㉝ Earthworm, woodlouse, millipede, etc. Any one.

㉞ 60% of sun's energy is not absorbed by the pigments within the chloroplasts because it is either the wrong wavelength, is reflected by the leaf surface, or is transmitted through the leaf without striking a chlorophyll molecule.

㉟ 6600 kJ is lost to decomposers, 1500 available to next level, so remainder must be lost heat during respiration, i.e. 15000 – 8100 = **6900 kJ**

㊱ 125 kJ enters level, minus 40 kJ lost to decomposers, minus 60 lost as heat leaves **25 kJ stored.**

㊲ 125 kJ enters level, 25 kJ is assimilated so efficiency = $\frac{25}{125} \times 100 = 20\%$

㊳ 1 = D, 2 = C, 3 = B, 4 = A

㊴ Mutualism is an interaction between organisms of two species from which both derive benefit whereas in commensalism one benefits but the other is not affected.

㊵ More trees would remove more carbon dioxide from the atmosphere via photosynthesis and incorporate the carbon into organic molecules, e.g. glucose.

㊶ Nitrogenase reduces nitrogen to ammonium ions.

㊷ It promotes nitrification and nitrogen fixation, but inhibits denitrification so leads to greater concentrations of soil nitrate. It also provides oxygen for aerobic respiration within the roots allowing for active transport of mineral ions.

㊸ Overhunting and collection; competition from introduced species; pollution; habitat destruction; natural selection (any three).

㊹ Legislation, e.g. CITES or EU Habitat Directive, and establishment of protected areas; Captive breeding programmes including gene banks; Education; Ecotourism.

㊺ Habitat loss; soil erosion; reduction in biodiversity; increased sedimentation and lowland flooding; reduction in soil quality and structure; climate change and reduced rainfall (any four).

㊻ Selective cutting harvests individual trees, whereas coppicing involves cutting each tree close to the base and later harvesting stems, so trees are not removed.

㊼ Imposing fish quotas; enforcing exclusion zones; restricting mesh size; enforcing fishing seasons; reduction in fleet size (any four).

㊽ Excretion eliminates waste produced by the body, whereas egestion removes waste not made by the body, e.g. undigested food.

㊾ Removal/excretion of nitrogenous waste and osmoregulation.

㊿ Any three from: water, glucose, salts (or named ions, e.g. sodium ions), amino acids, small proteins, e.g. HCG (< 68,000 rmm). Not in filtrate: large proteins (>68,000 rmm) e.g. albumin, cells.

(51) HCG must be <68,000 whilst other proteins are larger.

(52) In the medulla.

(53) The afferent arteriole is wider than the efferent arteriole, which causes increased blood pressure in the glomerulus.

(54) There is too much glucose present in the glomerular filtrate to be reabsorbed in the proximal convoluted tubule, so some remains. (Allow reference to limited number of transport proteins for glucose.)

(55) Osmoreceptors (located in the hypothalamus).

(56) Dissolved in In the plasma.

(57) Freshwater fish are able to excrete ammonia (which is more toxic than urea), due to the high availability of water to dilute it to a non-toxic concentration. Birds excrete uric acid as it is an adaptation for flight, reducing weight.

(58) A high proportion of their nephrons are juxtamedullary, i.e. they have long loops of Henle and produce small volumes of highly concentrated urine, because a higher ion concentration in the medulla can be created by the longer counter current multiplier. Their nephrons are referred to as being juxtamedullary nephrons, with the Bowman's capsule being located closer to the medulla, and loops of Henlé which penetrate deep into the medulla. Camels respire a significant mass of fats which liberate metabolic water.

(59) Motor neurones.

(60) Presence of myelin.

(61) Relay neurons.

(62)

Cnidarian nerve net	Mammalian nervous system
1 type of simple neurone	3 types of neurone (sensory, relay and motor)
Unmyelinated	Myelinated
Short, branched neurones	Long, unbranched neurones

(63) −70 mV

(64) A = IV, B = I, C = VI, D = III and II, E = V

(65) Another action potential cannot be generated until the resting potential is restored, which ensures a unidirectional impulse.

(66) Axon diameter and myelination (not temperature as mammals are warm blooded).

(67) Facilitated diffusion.

(68) They provide ATP for the synthesis AND exocytosis of acetylcholine.

Extra 3.2

Factor	Effect on TP	Effect on GP	Effect on RuBP
Light intensity		Decreasing light intensity means more GP because RuBP can be converted to GP but without ATP and reduced NADP GP will not be used up to make TP.	
Carbon dioxide concentration	As carbon dioxide increases TP increases, because more CO_2 is fixed, so more GP is made, so more TP.		As carbon dioxide increases RuBP decreases. Because more CO_2 is fixed, so more GP is made and more RuBP is used up.
Temperature	As temperature increases TP increases. But at high temperatures TP will decrease because the enzyme RuBisCO denatures and less carbon dioxide fixed, so less GP will be made and so less TP is made.		As temperature increases RuBP decreases because as the rate of enzyme action increases more RuBP is used up. When the RuBisCO denatures at high temperature less RuBP will be used up as CO_2 is not fixed.

Extra 3.3

Statement	Glycolysis	Link reaction	Krebs cycle	Electron transport chain
Is oxygen needed?	No	Yes	Yes	Yes
Is carbon dioxide produced?	No	Yes	Yes	No
Where does it take place	Cytoplasm	Matrix	Matrix	Cristae
Is FAD reduced?	No	No	Yes	No
Is NADH oxidised?	No	No	No	Yes

Extra 3.4

Dilution	Dilution factor	Number of colonies
1	10	Too many to count
2	100	>600
3	1000	59
4	10,000	9

Choose dilution plate no. 3.

$59 \times 1\,000 = 59\,000$ per $0.1\ cm^3$

$= 59\,000 \times 10 = 590\,000$ per cm^3 original culture

$= 590\,000 \times 25 = 14\,750\,000$ or 1.5×10^7 (1 dp).

Extra 3.7

a) Water is reabsorbed by osmosis, but urea is not reabsorbed (allow reference to a very small reabsorption of urea), so the concentration increases because the same mass of urea is present in a much lower volume of water.

b) The majority of sodium ions are reabsorbed after the proximal convoluted tubule. Not all are reabsorbed as some sodium ions are still present in the urine.

c) Not all the glucose is reabsorbed in the proximal convoluted tubule, so some remains in the filtrate which lowers the water potential. Cells become damaged as water is drawn out of the cells by osmosis.

Unit 4

① Seminiferous tubules.

② Seminal vesicle and prostate gland.

③ Provides ATP for movement.

④ a) Provide spermatozoa with nutrients and protect from the male's immune system.
b) Secrete testosterone.

⑤ The organism becomes independent of water because the gametes do not dehydrate, as the sperm is introduced directly into the female reproductive tract.

⑥ Digest cells of the corona radiata and zona pellucida allowing the sperm head to enter the oocyte.

⑦ Converts the zona pellucida into a fertilisation membrane preventing further sperm entry (polyspermy).

⑧ Exchange of gases, nutrients and waste; produces hormones

⑨ Acts as a shock absorber protecting the developing foetus; helps to maintain the foetus's body temperature.

⑩ Any two from

Insect-pollinated flowers	Wind-pollinated flowers
Large colourful petals, scent and nectar	Small, green and inconspicuous, no scent, petals usually absent
Anthers within the flower	Anthers hanging outside the flower so wind can blow pollen away
Stigma within the flower	Large feathery stigmas providing a large surface area to catch pollen grains
Small quantities of sticky pollen	Large quantities of small, smooth, light pollen

⑪ Ovary, stigma and style.

⑫ Lighter pollen can be carried more easily by the wind ensuring that pollination occurs.

⑬ The endosperm is formed from the triploid nucleus produced when the second male nuclei fuses with the diploid polar nucleus.

⑭ To provide an energy reserve for the developing embryo plant.

⑮ The diploid zygote divides by **mitosis** to form the embryonic plant.
The **endosperm** (the food reserve for the developing embryo) develops from the endosperm nucleus.
The **integuments** become the testa (seed coat) and the micropyle remains.
The fertilised ovule becomes the **seed**.
The fertilised ovary becomes the **fruit**.

⑯ Pollination is the transfer of pollen from the anther to the mature stigma of the same species of plant, whereas fertilisation is the fusion of the male and female gametes to produce a diploid zygote.

⑰ Optimum temperature for enzyme action, water for the mobilisation of enzymes and transport of products to growing points, and oxygen for aerobic respiration producing ATP for cellular processes such as protein synthesis.

⑱ To anchor the seedling and to absorb water to allow hydrolysis by enzymes and transport maltose and glucose to growing points.

⑲ A different form of the same gene that codes for a specific polypeptide, which is only expressed in the heterozygote, e.g. rr.

⑳ Where different genes are found on the same autosome chromosome and therefore cannot segregate independently.

㉑ Sex linkage is when a gene is carried by a sex chromosome so that a characteristic it encodes is seen predominately in one sex.

㉒ $X^H X^h$ $X^H y$

	X^H	Y
X^H	$X^H X^H$	$X^H Y$
X^h	$X^H X^h$	Y

1 $X^H X^H$ female normal
1 $X^h X^H$ female carrier
1 $X^H Y$ male normal
1 $X^h Y$ male haemophiliac
chance of having a male child with haemophilia is 1 in 4 or 0.25.

㉓ Normal male (he has a normal female child 7) and carrier female (normal female child 7, haemophiliac male 5).

㉔ Gene (point) and chromosome mutations.

㉕ Carcinogens.

㉖ It is discontinuous variation, as there are no intermediates, and is monogenic, i.e. controlled by one gene.

㉗ Any four:
  ▪ Point/gene or chromosome mutations.
  ▪ Crossing over during prophase I of meiosis.
  ▪ Independent assortment during metaphase I and II of meiosis.
  ▪ Random mating, i.e. that any organism can mate with another.
  ▪ Random fusion of gametes, i.e. the fertilisation of any male gamete with any female gamete.
  ▪ Environmental factors leading to epigenetic modifications.

㉘ It results in an increase in the frequency of advantageous alleles and decrease in the frequency of disadvantageous alleles within a population in a changing environment.

㉙ Due to differences in chromosome structure or number, chromosomes fail to pair during prophase I of meiosis and so gametes do not form.

㉚ Both involve the formation of new species from pre-existing ones, but allopatric speciation involves populations separated by a geographical barrier, whereas sympatric involves reproductive isolation of populations by other means, e.g. incompatible genitalia.

㉛ Electrophoresis separates molecules by size. It involves placing a voltage across an agarose gel and allowing charged molecules to migrate through the gel. Smaller molecules are able to move more easily than larger ones and so travel further.

㉜ To allow the primers to anneal (attach).

㉝ Because of the negative charge on the phosphate groups which are attracted to the positive electrode.

㉞ It confirms that the donor DNA has been inserted into the plasmid, because the marker gene will be inactivated.

㉟ Complementary or copy DNA is produced by reverse transcriptase from a mRNA template.

㊱ DNA ligase is a bacterial enzyme that joins sugar-phosphate backbones of two sections of DNA together: in this case the human insulin gene and bacterial plasmid DNA.

㊲ Mature mRNA does not contain introns.

㊳ Pollen may be transferred conferring herbicide resistance to other closely related plants.

㊴ A modified common cold virus (adenovirus) infects cells lining the respiratory tract, so could be used to introduce the normal CFTR allele to cells affected by cystic fibrosis.

㊵ The scaffold supports growth of a 3D tissue, and allows diffusion of nutrients and waste products.

㊶ Adult stem cells cannot differentiate into all types of cell whereas embryonic stem cells can.

㊷ Sterilisation kills all microbes and their spores, whilst disinfection kills most.

㊸ To ensure protective immune response is stimulated.

㊹ In active immunity, the individual produces antibodies in response to an infection or vaccination, whereas in passive immunity, the individual receives antibodies from either the placenta or breast milk or through an antibody injection. Active immunity gives longer lasting protection than passive due to the production of memory cells.

㊺ Osteoblasts secrete bone matrix around the cartilage; osteoclasts break down bone matrix.

㊻ Sarcomeres, myofibrils and muscle shorten. I band and H zone shorten.

㊼ Rich blood supply, large numbers of mitochondria, high myoglobin levels.

㊽ Vertebrarterial canals.

㊾ Secretes synovial fluid.

㊿ Reduces wear on articular surfaces.

(51) Rheumatoid arthritis is an autoimmune disorder, osteoarthritis is not. Rheumatoid arthritis has a genetic component whereas no genetic link has yet been identified in osteoarthritis. *Do not allow* osteoarthritis can affect any joint but rheumatoid arthritis more commonly affects the hands and wrists, because any joint can be affected.

(52) Pia mater.

(53) A = 5 B = 3 C = 4 D = 2 E = 1.

(54) a) temporal lobe b) occipital c) frontal lobe.

(55) Positron emission tomography (PET).

(56) Non-synaptic plasticity.

(57) Kineses are non-directional whereas in taxes the direction of the movement is related to the direction of the stimulus either towards or away from it.

(58) Operant conditioning.

(59) Classical conditioning.

(60) Animals occupy a home range but defend a territory.

Answers to Extra questions

4.1

a) A peak in FSH concentration at day 4 prior to rise in oestrogen levels.

b) Day 14 – peak in LH.

c) Oestrogen as it stimulates LH.

d) Progesterone; as it would inhibit FSH, so no oestrogen produced, resulting in no stimulation of LH production.

4.3

YYRr X yyRR

	YR	Yr
yR	YyRR Yellow round	YyRr Yellow round

all yellow round

4.4

Sufferers are homozygous recessive,

i.e. $q^2 = \frac{4.4}{10,000} = 0.00044$

$q = \sqrt{0.00044} = 0.0210$

$p = 1 - 0.0210 = 0.9790$

$2pq = 2 \times 0.9790 \times 0.0210 = 0.0411$

or 411 per 10,000 live births

4.5

a) i. Six fragments.

ii. Smallest fragment is 35 bases.

iii. 4400 – 3675 = 725 bases.

b) Hind 111 and Sna 1.

c) Two fragments: Bam 1 cuts as shown, Pst 1 does not cut DNA.

*GATTC***C|CTAGG***ATCGAAGTCGGGTTTAAA*
*CGAA***GGGATC|C***TAGCTTCAGCCCAAATTT*

GATTCC CTAGG*ATCGAAGTCGGGTTTAAA*
*CGAA*GGGATC *CTAGCTTCAGCCCAAATTT*

4B

Type of training	Effect of training	Advantage
Endurance training	Increase in number and size of mitochondria	More aerobic respiration possible
Endurance training	Capillary network increases	Increased blood supply to muscle results in more oxygen so more aerobic respiration.
Weight training	Increase in number of myofibrils and size of muscles	Increases strength
Endurance training	Increase in amount of myoglobin	Myoglobin is an oxygen store so more aerobic respiration
Weight training	Increase in tolerance to lactic acid	More anaerobic respiration possible

Answers to additional practice questions

1 a   Nucleotides;

   b) Photolysis/splitting of water
       replaces electrons lost from chlorophyll a in photosystem II
       provides {protons/H$^+$} to reduce NADP/for ATP synthesis

2 a) i)   Nitrification;

       ii) Denitrification;

   b) Atmospheric nitrogen turned into ammonium ions by nitrogen fixation / nitrogen fixing bacteria
       by *Rhizobium* in root nodules (of leguminous plants)
       by *Azotobacter* (free living) in soil

3 a) Any two but comparison needed.

Slow twitch	Fast twitch
Have more mitochondria	Have fewer mitochondria
Adapted for aerobic respiration	Adapted for anaerobic respiration
High resistance to fatigue	Lower resistance to fatigue
Continuous extended contraction	Generate short bursts of strength/speed
High density of capillaries	Low density of capillaries
Low density of myofibrils	High density of myofibrils
High concentration of myoglobin	Low concentration of myoglobin

   b)

Description	Reason
Capillary network density increases	More blood allows more oxygen, so more aerobic respiration
Increase in number/size of mitochondria	More aerobic respiration
Increase in concentration of myoglobin	Myoglobin is an oxygen store so more aerobic respiration
AVP, e.g. increase tolerance to lactate	E.g. more lactate can build up in the muscles so the athlete can run for longer

4. a) Gram-positive cocci

   b) Purple/Gram-positive bacteria have thicker cell wall made of peptidoglycan
       which takes up Gram stain/crystal violet
       Gram-negative bacteria have lipopolysaccharide layer that does not retain stain

5. a)

Statement	Glycolysis	Link reaction	Krebs cycle	Electron transport chain
Occurs in the mitochondrial matrix	✗	✓	✓	✗
ATP produced by substrate level phosphorylation	✓	✗	✓	✗
FAD reduced	✗	✗	✓	✗
Reduced NAD is oxidised	✗	✗	✗	✓

One mark per correct row

b) ATP phosphorylates glucose making it more reactive and easier to split into two molecules of triose phosphate.

6. a) Potential difference across the membrane of the neurone when a nerve impulse is not being generated.

the potential difference across the membrane of a neurone is negatively charged internally with respect to the outside/ is −70 mV (accept: −50 to −90 mV)
membrane is said to be polarised (any two)

b) Membrane is more permeable to K^+ / impermeable to Na^+ because some K^+ gates are OPEN (allows K^+ to pass out)

Na^+ gates are closed (prevents Na^+ entering);
Na^+/K^+ pump, actively transports 3K^+ in, 2Na^+ out.

7. *Many* mitochondria provide ATP for active transport

have microvilli/basal channels increasing surface area for diffusion
increased number of carrier proteins for facilitated diffusion/active transport/co-transport.

Reference to tight junctions preventing lateral transport.

8. a) Enzymes are produced which digest a path towards the female nucleus.
   b) Meiosis II in oogenesis only occurs after the sperm enters the secondary oocyte, whereas meiosis II in spermatogenesis follows directly after meiosis I.
   Four spermatids are produced from one germ cell in spermatogenesis, whereas only one ovum is produced in oogenesis with three polar bodies that subsequently disintegrate.
   All primary oocytes needed are produced before puberty, whereas primary spermatocytes are continually produced from puberty.

9. a) A gene is a length of DNA at a specific locus on a chromosome normally coding for a specific polypeptide, whereas an allele is a different form of the same gene with a different base sequence. A gene usually has two or more alleles.

   b) i) The law of independent assortment states that 'Each member of an allelic pair may combine randomly with either of another pair'.

   ii) If Mendel's second law applied you would expect to see green wrinkled and yellow smooth peas. Autosomal linkage must have occurred, i.e. that the genes for colour and texture were found on the same chromosome, so they were inherited together. This means that only one homologous pair of chromosomes is needed to accommodate all four alleles, which reduces the possible number of gamete types from four to two.

10. a) Any three

Continuous variation	Discontinuous variation
Range of phenotypes seen	Characteristics fit into distinct groups, there are no intermediates
Controlled by many genes (polygenic)	Usually controlled by one gene with two or more alleles (monogenic)
Follows a 'normal' distribution of phenotypes	Does not follow a 'normal' distribution
Environmental factors have a major influence, e.g. diet on weight	Environmental factors have little influence, e.g. diet has no effect on blood group

b) Organisms overproduce offspring, there is a large variation of genotypes in population.

Changes to environmental conditions bring new selection pressures through competition/ predation/ disease.

Only those individuals with beneficial alleles conferring a beneficial phenotype have a selective advantage, e.g. light fur in sandy environment, and dark fur in the rocky environment, so therefore are more likely to survive due to their camouflage, and reproduce.

Offspring are likely to inherit the beneficial alleles, so therefore the beneficial allele frequency increases within the gene pool.

11. a) The gel is made from agarose, which contains pores in its matrix. DNA samples are loaded at one end and a voltage is applied across the gel. DNA is attracted to the positive electrode due to the negative charge present on the phosphate groups. Smaller fragments migrate more easily through the pores in the gel and so travel further than large fragments in the same time.

b) Identify the allele for normal haemoglobin, and extract normal haemoglobin mRNA.

Use reverse transcriptase to produce cDNA from mRNA template.

Insert into plasmid or virus, and inject into bone marrow of patient.

12. Main treatment is by haemodialysis in which blood passes into a dialyser with a selectively permeable membrane and dialysis fluid flows in opposite direction to the direction of blood flow (referred to as counter current flow). Dialysis fluid has same water potential and glucose concentration as normal blood, so urea, excess water and salts diffuse out into dialysis fluid. Peritoneal dialysis uses the peritoneum which acts as a filter once dialysis fluid is passed into the abdominal cavity through a catheter.

The fluid is drained off after a period of time, removing waste, e.g. urea. In severe cases a kidney transplant can be performed which involves surgically transplanting a kidney from a donor. The donor must be a close tissue type and blood group match to recipient. Immuno-suppressant drugs must be used to reduce chance of rejection.

13. a) Humoral

b) AIDS is end stage HIV infection so helper T cells are destroyed resulting in fewer helper T cells so they can't stimulate B lymphocytes to produce antibodies.

14. a) The medulla oblongata controls heart rate and ventilation, and the cerebellum is involved with the maintenance of posture.

b) Functional magnetic resonance imaging (fMRI) is a technique for examining activity of brain tissue in real time as opposed to structure. It uses a radio wave pulse (in addition to a magnetic field) which interacts differently with haemoglobin and oxyhaemoglobin showing areas where there is greater oxygen demand and therefore more aerobic respiration.

Positron emission tomography (PET) is a neuroimaging technique which involves the injection of a small amount of radioactive isotope fluorodeoxyglucose, which is taken up into active cells and then emits a positron as it decays. This is detected by the scanner and shows where glucose is being actively respired in the brain. The isotope used has a short half-life and so is quickly eliminated from the body.

Not computerised tomography (CT) scans or Magnetic resonance imaging (MRI) as they show structure not function in the brain.

15. i) $\dfrac{549 - 443}{443} \times 100$

= 23.9% acc. 24%

ii) **Healthy person**
myelin sheath prevents action potential / action potential only forms at nodes;
action potential 'jumps' from node to node /saltatory conduction;
greatly increasing nerve conductance speed.

in M.S. sufferers
(demyelination occurs in M.S. sufferers)
nerve conductance in motor neurones is slower;
lack of myelination prevents saltatory conduction;
depolarisation occurs along whole length of neurone.

iii) sensory neurones/optic nerve affected/lose myelin

Index